**Lectures in Mathematics
ETH Zürich**
Department of Mathematics
Research Institute of Mathematics

Managing Editor:
Helmut Hofer

Leon Simon
Theorems on Regularity and Singularity of Energy Minimizing Maps

based on lecture notes by
Norbert Hungerbühler

Birkhäuser Verlag
Basel · Boston · Berlin

Author's address:

Leon Simon
Department of Mathematics
Stanford University
Stanford, CA 94305
USA

1991 Mathematics Subject Classification 58C27, 58E20

A CIP catalogue record for this book is available from the Library of Congress, Washington D.C., USA

Deutsche Bibliothek Cataloging-in-Publication Data
Simon, Leon:
Theorems on regularity and singularity of energy minimizing maps : based on lecture notes by Norbert Hungerbühler / Leon Simon. – Basel ; Boston ; Berlin : Birkhäuser, 1996
   (Lectures in mathematics)
   ISBN 3-7643-5397-X (Basel ...)
   ISBN 0-8176-5397-X (Boston)

This work is subject to copyright. All rights are reserved, whether the whole or part of the material is concerned, specifically the rights of translation, reprinting, re-use of illustrations, recitation, broadcasting, reproduction on microfilms or in other ways, and storage in data banks. For any kind of use permission of the copyright owner must be obtained.

© 1996 Birkhäuser Verlag, P.O. Box 133, CH-4010 Basel, Switzerland
Printed on acid-free paper produced of chlorine-free pulp. TCP ∞
Printed in Germany
ISBN 3-7643-5397-X
ISBN 0-8176-5397-X

9 8 7 6 5 4 3 2 1

Contents

Preface vii

1 Analytic Preliminaries

   1.1 Hölder Continuity 1
   1.2 Smoothing 4
   1.3 Functions with L^2 Gradient 5
   1.4 Harmonic Functions 8
   1.5 Weakly Harmonic Functions 10
   1.6 Harmonic Approximation Lemma 10
   1.7 Elliptic regularity 11
   1.8 A Technical Regularity Lemma 13

2 Regularity Theory for Harmonic Maps

   2.1 Definition of Energy Minimizing Maps 19
   2.2 The Variational Equations 20
   2.3 The ε-Regularity Theorem 22
   2.4 The Monotonicity Formula 23
   2.5 The Density Function 24
   2.6 A Lemma of Luckhaus 25
   2.7 Corollaries of Luckhaus' Lemma 26
   2.8 Proof of the Reverse Poincaré Inequality 29
   2.9 The Compactness Theorem 32
   2.10 Corollaries of the ε-Regularity Theorem 35
   2.11 Remark on Upper Semicontinuity of the Density $\Theta_u(y)$ 37
   2.12 Appendix to Chapter 2 37
         2.12.1 Absolute Continuity Properties of Functions in $W^{1,2}$ 37
         2.12.2 Proof of Luckhaus' Lemma (Lemma 1 of Section 2.6) 38
         2.12.3 Nearest point projection 42
         2.12.4 Proof of the ε-regularity theorem in case $n = 2$ 46

3 Approximation Properties of the Singular Set

- 3.1 Definition of Tangent Map 51
- 3.2 Properties of Tangent Maps 52
- 3.3 Properties of Homogeneous Degree Zero Minimizers 52
- 3.4 Further Properties of $\operatorname{sing} u$ 54
- 3.5 Definition of Top-dimensional Part of the Singular Set 58
- 3.6 Homogeneous Degree Zero φ with $\dim S(\varphi) = n - 3$ 58
- 3.7 The Geometric Picture Near Points of $\operatorname{sing}_* u$ 59
- 3.8 Consequences of Uniqueness of Tangent Maps 61
- 3.9 Approximation properties of subsets of $\mathbb{R}^n$ 62
- 3.10 Uniqueness of Tangent maps with isolated singularities 67
- 3.11 Functionals on vector bundles 72
- 3.12 The Liapunov-Schmidt Reduction 74
- 3.13 The Łojasiewicz Inequality for $\mathcal{F}$ 78
- 3.14 Łojasiewicz for the Energy functional on S^{n-1} 80
- 3.15 Proof of Theorem 1 of Section 3.10 82
- 3.16 Appendix to Chapter 3 87
  - 3.16.1 The Liapunov-Schmidt Reduction in a Finite Dimensional Setting 87

4 Rectifiability of the Singular Set

- 4.1 Statement of Main Theorems 91
- 4.2 A general rectifiability lemma. 92
- 4.3 Gap Measures on Subsets of $\mathbb{R}^n$ 104
- 4.4 Energy Estimates 110
- 4.5 L^2 estimates 120
- 4.6 The deviation function ψ 129
- 4.7 Proof of Theorems 1, 2 of Section 4.1 135
- 4.8 The case when Ω has arbitrary Riemannian metric 143

Bibliography 147

Index 150

Preface

This monograph was started at the ETH during the Spring Semester of 1993, when the author was presenting a Nachdiplom course on regularity and singularity of energy minimizing maps between Riemannian manifolds.

The aim here (as was the aim of the lectures) is to give an essentially self-contained introduction to the basic regularity theory for energy minimizing maps, including recent developments concerning the structure of the singular set and asymptotics on approach to the singular set.

It is not assumed that the reader has prior specialized knowledge in partial differential equations or the geometric calculus of variations; a good general background in mathematical analysis would be adequate preparation.

I want to thank the ETH for the opportunity to present this course in such congenial surroundings. Most of all I have to thank Norbert Hungerbühler for his patience, efficiency and expertise in producing the beautiful LaTeX 2_ε files, complete with figures, from the lectures and the rather scrappy additional material that I from time to time gave to him.

The support of National Science Foundation Grants DMS–9504456 and DMS–9207704 at Stanford University during part of the preparation of this monograph and during part of the research described in Chapter 4 is also acknowledged.

Stanford, August 1995   Leon Simon

Chapter 1

Analytic Preliminaries

1.1 Hölder Continuity

If $\Omega \subset \mathbb{R}^n$ is open and if $\alpha \in (0,1]$, $u : \Omega \to \mathbb{R}$ is said to be *Hölder continuous with exponent α on Ω* if there is a constant C such that

(i) $$|u(x) - u(y)| \le C|x-y|^\alpha$$

for every $x, y \in \Omega$. u is called *locally Hölder continuous* on Ω with exponent α if it is Hölder continuous on each $\widetilde{\Omega} \subset\subset \Omega$, where $\widetilde{\Omega} \subset\subset \Omega$ means that the closure of $\widetilde{\Omega}$ is a compact subset of Ω. For the Hölder continuous functions on Ω we have a semi-norm, called the Hölder coefficient of u, defined by

$$[u]_{\alpha;\Omega} := \sup_{x \ne y,\, x,y \in \Omega} \frac{|u(x) - u(y)|}{|x-y|^\alpha}.$$

This Hölder coefficient is not a norm because it is zero for any constant function; it is characterized by being the smallest constant C such that (i) holds for every $x, y \in \Omega$. We will write $C^{0,\alpha}(\overline{\Omega})$ for the bounded Hölder continuous functions on Ω. The space $C^{0,\alpha}(\overline{\Omega})$ becomes a Banach space in the norm

(ii) $$|u|_{0,\alpha;\Omega} := \|u\|_{L^\infty(\Omega)} + [u]_{\alpha;\Omega}.$$

This can easily be checked with the aid of the Arzela-Ascoli Lemma. Finally, we call u *Lipschitz continuous* if u is Hölder continuous with exponent 1.

Hölder continuity turns out to be of fundamental importance in geometric analysis and PDE. We mention here two facts about Hölder continuity which give an initial hint as to why this might be so:

(a) Scaling: If $|u(x) - u(y)| \le \beta |x-y|^\alpha$ for every $x, y \in \Omega$ and if for given $R > 0$ we define the scaled function $\widetilde{u}(x) = R^{-\alpha} u(Rx)$ for $x \in \widetilde{\Omega} := \{R^{-1} y : y \in \Omega\}$, then $|\widetilde{u}(x) - \widetilde{u}(y)| \le \beta|x-y|^\alpha$ for every $x, y \in \widetilde{\Omega}$. In fact we evidently have

$[u]_{\alpha;\Omega} = [\widetilde{u}]_{\alpha;\widetilde{\Omega}}$. Notice that other kinds of continuity do not have such nice "scaling invariance" properties.

(b) Dyadic decay of oscillation: For real valued functions u defined on $\Omega \subset \mathbb{R}^n$ the *oscillation* is defined as

$$\operatorname{osc}_\Omega u := \sup_{x\in\Omega} u(x) - \inf_{x\in\Omega} u(x).$$

If $u : B_R(x_0) \to \mathbb{R}$ with $\operatorname{osc}_{B_R(y)} u < \infty$ and if there is a fixed $\theta \in (0, \tfrac{1}{2})$ such that

(ii) $$\operatorname{osc}_{B_{\theta\rho}(y)} u \leq \frac{1}{2} \operatorname{osc}_{B_\rho(y)} u$$

for every $y \in B_{R/2}(x_0)$ and every $\rho \leq \tfrac{R}{2}$, then $u \in C^{0,\alpha}(\overline{B}_{R/2}(x_0))$ with $\alpha = -\frac{\log 2}{\log \theta}$, and moreover

$$[u]_{\alpha;B_{R/2}(x_0)} \leq R^{-\alpha} C_\theta \operatorname{osc}_{B_R(x_0)} u.$$

Proof: By induction we get from (ii) for $k = 0, 1, 2, \ldots$ the estimate

(1) $$\operatorname{osc}_{B_{\theta^k \rho}(y)} u \leq 2^{-k} \operatorname{osc}_{B_\rho(y)} u$$

valid for all $y \in B_{R/2}(x_0)$ and all $\rho \leq \tfrac{R}{2}$. Now, for x with $r := |x-y| < \tfrac{R}{2}$ we can choose an integer k such that $\theta^{k+1} \leq \tfrac{2r}{R} < \theta^k$. Using (1) with $\rho = R/2$, we get

$$\begin{aligned} |u(x) - u(y)| &\leq 2^{-k} \operatorname{osc}_{B_{R/2}(y)} u \leq 2^{-k} \operatorname{osc}_{B_R(x_0)} u \\ &= \theta^{k\alpha} \operatorname{osc}_{B_R(x_0)} u \leq \theta^{-\alpha} \left(\frac{2|x-y|}{R}\right)^\alpha \operatorname{osc}_{B_R(x_0)} u \end{aligned}$$

provided α is chosen so that $\theta^\alpha = \tfrac{1}{2}$. Thus the claim is established. $\square$

The following result of Campanato is a nice characterization of those L^2 functions which are Hölder continuous on a ball in $\mathbb{R}^n$:

Lemma 1 *Suppose $u \in L^2(B_{2R}(x_0))$, $\alpha \in (0,1]$, $\beta > 0$ and*

(iii) $$\inf_{\lambda \in \mathbb{R}} \rho^{-n} \int_{B_\rho(y)} |u - \lambda|^2 \leq \beta^2 \left(\frac{\rho}{R}\right)^{2\alpha}$$

for every ball $B_\rho(y)$ such that $y \in B_R(x_0)$ and $\rho \leq R$. Then there is a Hölder continuous representative $\overline{u}$ for the L^2-class of u with

$$|\overline{u}(x) - \overline{u}(y)| \leq C_{n,\alpha} \beta \left(\frac{|x-y|}{R}\right)^\alpha, \quad \forall\, x, y \in B_R(x_0),$$

where $C_{n,\alpha}$ depends only on n and α.

1.1. Hölder Continuity

Remark: Notice that, since $\int_{B_\rho(y)} |u - \lambda|^2 = \int_{B_\rho(y)} (u^2 - 2\lambda u + \lambda^2)$, it is easy to check that the infimum on the left of (iii) is attained when, and only when, λ has the average value $\lambda_{y,\rho} := (\omega_n \rho^n)^{-1} \int_{B_\rho(y)} u$, where ω_n is the volume of the unit ball in $\mathbb{R}^n$. We make frequent use of this in the sequel.

Proof: First note that

$$(1) \quad \left(\frac{\rho}{2}\right)^{-n} \int_{B_{\rho/2}(y)} |u - \lambda_{y,\rho}|^2 \leq 2^n \rho^{-n} \int_{B_\rho(y)} |u - \lambda_{y,\rho}|^2 \leq 2^n \beta^2 \left(\frac{\rho}{R}\right)^{2\alpha},$$

where $\lambda_{y,\rho}$ is the average of u over $B_\rho(y)$ as in the above remark. Using the given inequality with $\rho/2$ in place of ρ, we also have

$$(2) \quad \left(\frac{\rho}{2}\right)^{-n} \int_{B_{\rho/2}(y)} |u - \lambda_{y,\rho/2}|^2 \leq 2^n \beta^2 \left(\frac{\rho}{R}\right)^{2\alpha}.$$

Adding (1) and (2) and using the squared triangle inequality $|a - b|^2 \leq 2|a - c|^2 + 2|b - c|^2$, we conclude that

$$(3) \quad |\lambda_{y,\rho} - \lambda_{y,\rho/2}| \leq 2^n \omega_n^{-1/2} \beta \left(\frac{\rho}{R}\right)^\alpha$$

provided that $\rho \leq R$ and $y \in B_R(x_0)$. (ω_n = the volume of the unit ball in $\mathbb{R}^n$.)

Now for any integer $\nu \in \{0, 1, 2, \ldots\}$ we can choose $\rho = 2^{-\nu}R$, whereupon (3) gives

$$(4) \quad |\lambda_{y,R/2^{\nu+1}} - \lambda_{y,R/2^\nu}| \leq 2^n \omega_n^{-1/2} \beta 2^{-\nu\alpha}.$$

Since $2^{-\nu\alpha}$ is the νth term of a convergent geometric series, we see that the series s defined by $s = \sum_{\nu=0}^\infty (\lambda_{y,R/2^{\nu+1}} - \lambda_{y,R/2^\nu})$ is absolutely convergent. But the jth partial sum s_j is just $\lambda_{y,2^{-j}R} - \lambda_{y,R}$, so we have $\lim_{\nu \to \infty} \lambda_{y,2^{-\nu}R}$ exists, and we denote this limit by λ_y. Using (4) again we see also that

$$(5) \quad |\lambda_{y,2^{-\nu}R} - \lambda_y| \leq \sum_{j=\nu}^\infty |\lambda_{y,R/2^{j+1}} - \lambda_{y,R/2^j}| \leq C\beta 2^{-\nu\alpha},$$

where C depends only on n and α. Then combining (1) (with $\rho = 2^{-\nu}R$) and (5), and using the squared triangle inequality again, we conclude

$$(6) \quad \rho^{-n} \int_{B_\rho(y)} |u - \lambda_y|^2 \leq C\beta^2 \left(\frac{\rho}{R}\right)^{2\alpha},$$

for $\rho = 2^{-\nu}R$, $\nu = 0, 1, \ldots$. On the other hand for *any* $\rho \in (0, R]$ there is an integer $\nu \geq 0$ such that $2^{-\nu-1}R < \rho \leq 2^{-\nu}R$, and it evidently follows (since $B_{2^{-\nu}R}(y) \supset B_\rho(y)$ for such ν) that (6) holds (with $2^{n+2\alpha}C$ in place of C) for *every* $\rho \leq R/2$.

Now take any pair of points $y, z \in B_R(x_0)$ with $|y - z| \leq R/4$, and apply (6) with $\rho = 2|y - z|$ on balls with centers y, z and add the resultant inequalities. Since $B_{\rho/2}(\frac{1}{2}(y+z)) \subset B_\rho(y) \cap B_\rho(z)$ this gives

$$\rho^{-n} \int_{B_{\rho/2}(\frac{1}{2}(y+z))} (|u - \lambda_y|^2 + |u - \lambda_z|^2) \leq 2C\beta^2 \left(\frac{\rho}{R}\right)^{2\alpha}.$$

Since $|\lambda_y - \lambda_z|^2 \leq 2|u - \lambda_y|^2 + 2|u - \lambda_z|^2$, this in turn gives

(7) $$|\lambda_y - \lambda_z| \leq 2C\beta(\rho/R)^\alpha = 2^{1+\alpha}C\beta\left(\frac{|y-z|}{R}\right)^\alpha.$$

Now for any pair of points $y, z \in B_R(x_0)$ we can pick points $z_0 = y, \ldots, z_8 = z$ on the line segment joining y, z such that $|z_i - z_{i-1}| \leq R/4$, and applying (7) to each of the pairs z_{i-1}, z_i and adding, we deduce finally that

(8) $$|\lambda_y - \lambda_z| \leq C\beta 2\left(\frac{|y-z|}{R}\right)^\alpha, \quad \forall y, z \in B_R(x_0).$$

On the other hand by letting $\rho \downarrow 0$ in (6) and using the Lebesgue Lemma, we have $\lambda_y = u(y)$ for almost all $y \in B_R(x_0)$, so the proof is complete (because then $\bar{u}(y) \equiv \lambda_y$ is a representative for the L^2 class of u which by (8) satisfies the required Hölder estimate). □

1.2 Smoothing

Consider a function $\varphi \in C_c^\infty(\mathbb{R}^n)$ having compact support in $B_1(0)$ with the properties $\varphi \geq 0$ and $\int_{B_1(0)} \varphi = 1$. For $\sigma > 0$ we define the mollifiers

$$\varphi^{(\sigma)}(x) := \sigma^{-n}\varphi\left(\frac{x}{\sigma}\right).$$

Note that $\varphi^{(\sigma)}$ has compact support in $B_\sigma(0)$ and $\int_{B_\sigma(0)} \varphi^{(\sigma)} = 1$. Now we can use the mollifiers $\varphi^{(\sigma)}$ to smooth a function $u \in L^1_{\text{loc}}(\Omega)$ ($\Omega \subset \mathbb{R}^n$). Let

$$\Omega_\sigma := \{x \in \Omega : \text{dist}(x, \partial\Omega) > \sigma\}$$

and for $x \in \Omega_\sigma$

$$u_\sigma(x) := \int_\Omega u(y)\varphi^{(\sigma)}(x-y)\,dy = (u * \varphi^{(\sigma)})(x).$$

u_σ is evidently smooth on Ω_σ. In fact, by virtue of the usual "differentiation under the integral" lemma, $D^\alpha u_\sigma$ is given explicitly by

$$D^\alpha u_\sigma(x) = \int_\Omega u(y) D^\alpha \varphi^{(\sigma)}(x-y)\,dy$$

for all $x \in \Omega_\sigma$. Here, we use the multi-index notation, i.e. $\alpha = (\alpha_1, \ldots, \alpha_n) \in \mathbb{Z}_+^n$ ($\mathbb{Z}_+ = \{0, 1, 2, \ldots\}$) with $|\alpha| = \sum_{j=1}^n \alpha_j$ and $D^\alpha = \frac{\partial^{|\alpha|}}{\partial x_1^{\alpha_1} \ldots \partial x_n^{\alpha_n}}$. Hence, we have that $u_\sigma \in C^\infty(\Omega_\sigma)$. Furthermore, the following approximation statements hold for $\sigma \to 0$:

(i) $u_\sigma \to u$ a.e. on Ω
(ii) $u_\sigma \to u$ in $L^1(K)$ for all compact sets $K \subset \Omega$
(iii) $u_\sigma \to u$ in $L^p(K)$ for all compact sets $K \subset \Omega$ if $u \in L^p_{\text{loc}}(\Omega)$
(iv) $u_\sigma \to u$ uniformly on compact sets K if u is continuous.

(i) follows from the Lebesgue Lemma, (iv) is evident from the definition, and (ii), (iii) follow from (iv) together with the fact that the continuous functions are dense in L^p.

1.3 Functions with L^2 Gradient

Recall that we say $u \in L^2(\Omega)$ has a *gradient in* $L^2(\Omega)$ if there exist functions $r_1, \ldots, r_n \in L^2(\Omega)$ such that

(i) $$\int_\Omega \varphi r_i = - \int_\Omega u D_i \varphi, \quad \forall \varphi \in C_c^\infty(\Omega).$$

Of course if such functions exist they are unique; further if $u \in C^1(\Omega)$ then (using integration by parts) such an identity holds with r_j equal to the usual partial derivative $D_j u$, so this notion of L^2 gradient really does generalize the classical notion of partial derivatives, and we call r_j the L^2 *weak derivatives of* u. We denote them, when they exist, simply by $D_j u$. In this case the identity (i) takes the familiar form

(ii) $$\int_\Omega \varphi D_i u = - \int_\Omega u D_i \varphi, \quad \forall \varphi \in C_c^\infty(\Omega).$$

Using the completeness of the space $L^2(\Omega)$, it is easy to check that the set of all functions $u \in L^2(\Omega)$ having L^2 gradient, equipped with the inner product

$$\langle u, v \rangle := \langle u, v \rangle_{L^2(\Omega)} + \sum_{j=1}^n \langle D_j u, D_j v \rangle_{L^2(\Omega)},$$

becomes a Hilbert space. This space will henceforth be denoted $W^{1,2}(\Omega)$ and the inner product norm will be denoted by $\| \cdot \|_{W^{1,2}(\Omega)}$. It is an example of the more general Sobolev spaces $W^{k,p}(\Omega)$.

Smoothing gives a nice relation between classical partial derivatives and L^2 partial derivatives as follows: If $u \in W^{1,2}(\Omega)$ then

(iii) $$D_j u_\sigma = (D_j u)_\sigma, \quad j = 1, \ldots, n, \; \sigma > 0,$$

holds on Ω_σ as one easily checks by using the definition (ii), keeping in mind that, for fixed $x \in \Omega_\sigma$, $\varphi^{(\sigma)}(x - y)$ is a $C_c^\infty(\Omega)$ function of y. Note that on the left hand side of (iii) D_j is the classical derivative of a smooth function, whereas on the right hand side D_j is the weak derivative.

We now present some of the important facts about $W^{1,2}$ functions.

First recall that an open subset $\Omega \subset \mathbb{R}^n$ is said to be *Lipschitz* if for each $x_0 \in \partial\Omega$ there exists $R > 0$ and a bijective function $\Phi_{x_0} : B_R(x_0) \to U \subset \mathbb{R}^n$ with the following properties

1. Φ_{x_0} is bi-Lipschitz, i.e. Φ_{x_0} and $\Phi_{x_0}^{-1}$ are Lipschitz functions.

2. If $U_+ := U \cap \mathbb{R}_+$ with $\mathbb{R}_+ := \{x = (x^1, \ldots, x^n) : x^n > 0\}$ then $\Phi_{x_0}(B_R(x_0) \cap \Omega) = U_+$.

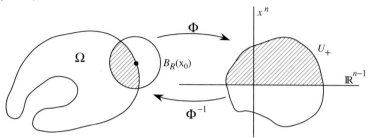

Thus in particular $\Phi_{x_0}(\partial\Omega \cap B_R(x_0)) = U \cap (\{0\} \times \mathbb{R}^{n-1})$, so that Φ_{x_0} "flattens" the boundary near x_0. A cube in $\mathbb{R}^n$ is an example of a Lipschitz domain (edges are allowed).

Now, if Ω is a bounded Lipschitz domain it is an exercise (based on the use of the local flattening transformations above together with even reflection across the boundary of the half-space) to show that there exists a continuous linear extension operator $E: W^{1,2}(\Omega) \to W^{1,2}(\mathbb{R}^n), u \mapsto \tilde{u}$: for every $u \in W^{1,2}(\Omega)$ there exists an extension $\tilde{u} \in W^{1,2}(\mathbb{R}^n)$ of u with $\|\tilde{u}\|_{W^{1,2}(\mathbb{R}^n)} \leq C_\Omega \|u\|_{W^{1,2}(\Omega)}$ where C_Ω depends only on Ω. Moreover, it is possible to arrange such an extension E such that $\{x \in \mathbb{R}^n : \tilde{u}(x) \neq 0\} \subset \{x \in \mathbb{R}^n : \text{dist}(x, \overline{\Omega}) \leq 1\}$. See e.g. [Ad75] or [GT83, Theorem 7.25] for a detailed proof of these facts.

Lemma 1 (Rellich Compactness Lemma) *Suppose Ω is a bounded Lipschitz domain in $\mathbb{R}^n$ and u_k is a sequence of $W^{1,2}(\Omega)$ with $\sup_k \|u_k\|_{W^{1,2}(\Omega)} < \infty$. Then there is a subsequence $u_{k'}$ and $u \in W^{1,2}(\Omega)$ such that*

(a) $u_{k'} \rightharpoonup u$ *weakly in* $W^{1,2}(\Omega)$,

(b) $u_{k'} \to u$ *strongly in* $L^2(\Omega)$,

(c) $\int_\Omega |Du|^2 \leq \liminf_{k' \to \infty} \int_\Omega |Du_{k'}|^2$.

Remark: In other words, the lemma claims that a bounded set in $W^{1,2}(\Omega)$ is precompact in $L^2(\Omega)$. In fact the same is true in $L^p(\Omega)$ for all $p < 2^* = \frac{2n}{n-2}$. See e.g. [Ma83] or [GT83, Chapter 7].

Proof: First note that the weak convergence of a subsequence of u_k, Du_k in $L^2(\Omega)$ is a consequence of the general weak compactness of the unit ball in a Hilbert space.

Next note that by the remarks preceding the lemma we can extend u_k to $\tilde{u}_k \in W^{1,2}(\mathbb{R}^n)$ with support contained in a fixed ball B (independent of k) and with $\sup_k \|\tilde{u}_k\|_{W^{1,2}(\mathbb{R}^n)} < \infty$. Now the rest of the proof involves combining the two facts

1.3. Functions with L^2 Gradient

that (a) the functions $(\widetilde{u}_k)_\sigma$ are bounded and have bounded derivatives on $\mathbb{R}^n$ with bounds depending on σ but not on k (indeed by using the definition of $(u_k)_\sigma$ we have $\sup |D(u_k)_\sigma| \leq C\sigma^{-1} \|u_k\|_{L^1} \leq C\sigma^{-1} \|u_k\|_{L^2} \leq C\sigma^{-1}$ with C independent of k), and (b) $\|\widetilde{u}_k - (\widetilde{u}_k)_\sigma\|_{L^2(\mathbb{R}^n)} \leq C\sigma \ \forall \sigma > 0$, with C independent of both k and σ. (See [GT83, p. 162] for the precise argument, which involves simply using the definition of $(\widetilde{u}_k)_\sigma$ and calculus.) (a) enables us to apply the Arzela-Ascoli lemma to $(u_k)_\sigma$ to find a uniformly convergent subsequence $(u_{k_j})_\sigma$ (depending on σ) which then by (b) (after dropping finitely many terms at the start of the subsequence) gives $\|\widetilde{u}_{k_j} - \widetilde{u}_{k_\ell}\|_{L^2(\mathbb{R}^n)} \leq C\sigma$ for each j, ℓ. Since this can be repeated with each $\sigma = 2^{-i}$, choosing a subsequence depending on i for each i, and since the $(i+1)^{\text{st}}$ subsequence can be chosen as a subsequence of the ith, by taking a diagonal sequence it then follows that there is a subsequence which is Cauchy with respect to the L^2-norm. By completeness of L^2 this proves the result *(b)*.

Finally, the fact that a Hilbert space norm is weakly lower semicontinuous implies that we have $\|u\|_{W^{1,2}(\Omega)} \leq \liminf_{k'\to\infty} \|u_{k'}\|_{W^{1,2}(\Omega)}$, which implies *(c)*. □

We shall also need the following Poincaré Inequality:

Lemma 2 (Poincaré Inequality) *Suppose Ω is a bounded and connected Lipschitz domain in $\mathbb{R}^n$. Then there exists a constant C_Ω depending only on the domain Ω such that for every function $u \in W^{1,2}(\Omega)$ there holds*

(iv) $$\int_\Omega |u - \lambda|^2 \leq C_\Omega \int_\Omega |Du|^2,$$

where $\lambda = |\Omega|^{-1} \int_\Omega u$.

Proof: There are various proofs of the Poincaré inequality; one nice way to prove it is to use the above Rellich compactness Theorem. Suppose the assertion is false: Then for each k, $k = 1, 2, \ldots$, there exist functions $u_k \in W^{1,2}(\Omega)$ such that (iv) fails with $C_\Omega = k$:

(*) $$\int_\Omega |Du_k|^2 < \frac{1}{k} \inf_{\lambda \in \mathbb{R}} \int_\Omega |u_k - \lambda|^2 \equiv \frac{1}{k} \int_\Omega |u_k - \lambda_k|^2$$

with $\lambda_k = \frac{1}{|\Omega|} \int_\Omega u_k$. Defining

$$v_k := \frac{u_k - \lambda_k}{\|u_k - \lambda_k\|_{L^2(\Omega)}}$$

we find that $\|v_k\|_{L^2(\Omega)} = 1$ and $\|Dv_k\|_{L^2(\Omega)} \leq \frac{1}{\sqrt{k}}$ and hence that $\{v_k\}$ is a bounded sequence in $W^{1,2}(\Omega)$. According to Rellich's compactness result (Lemma 1 above) there exists $v \in W^{1,2}(\Omega)$ such that $v_{k'} \to v$ strongly in $L^2(\Omega)$ for a subsequence $\{v_{k'}\}$. Further, since $\int_\Omega v_k = 0$ for all k, we conclude that $\int_\Omega v = 0$, and moreover (by Lemma 1) $Du_{k'} \to Dv$ (weakly in L^2) and $\|Dv\|_{L^2(\Omega)} \leq \liminf_{k'\to\infty} \|Dv_{k'}\|_{L^2(\Omega)} = 0$. Hence, $Dv = 0$ a.e. on Ω. Now it is easy to check that since Ω was supposed to be connected, $Dv = 0$ a.e. implies that v is constant. (For example, $Dv_\sigma = (Dv)_\sigma = 0$ on Ω_σ, so v_σ is constant on any connected $\tilde\Omega \subset \Omega_\sigma$, so we conclude $v =$ constant in Ω by letting $\sigma \downarrow 0$.) Now $\|v\|_{L^2(\Omega)} = 1$ clearly contradicts the fact that v has mean value zero. □

Remarks: By using the special case when Ω is the unit ball $B_1(0)$ and by changing scale $x \to Rx$, one can find the explicit dependence of the constant $C_{B_R(x_0)}$ on R; namely,

$$R^{-n}\int_{B_R(x_0)} |u - \lambda|^2 \leq CR^{2-n}\int_{B_R(x_0)} |Du|^2$$

for every $u \in W^{1,2}(B_R(x_0))$, where C is a constant depending only on n (and not on R).

Finally we want to state and prove Morrey's Lemma.

Lemma 3 (Morrey's Lemma) *Suppose $u \in W^{1,2}(B_R(x_0))$, $\beta > 0$, $\alpha \in (0, 1]$ are constants, and*

$$\rho^{2-n}\int_{B_\rho(y)} |Du|^2 \leq \beta^2 \left(\frac{\rho}{R}\right)^{2\alpha}, \qquad \forall\, y \in B_{R/2}(x_0),\, \rho \in (0, \tfrac{R}{2}].$$

Then $u \in C^{0,\alpha}(\overline{B}_{R/2}(x_0))$, and in fact

$$|u(x) - u(y)| \leq C\beta \left(\frac{|x-y|}{R}\right)^\alpha, \qquad \forall\, x, y \in B_{R/2}(x_0),$$

where C depends only on n.

Proof: Let $\lambda_{y,\rho} = (\omega_n \rho^n)^{-1}\int_{B_\rho(y)} u$. The Poincaré inequality gives

$$\rho^{-n}\int_{B_\rho(y)} |u - \lambda_{y,\rho}|^2 \leq C\rho^{2-n}\int_{B_\rho(y)} |Du|^2 \leq C\beta^2 \left(\frac{\rho}{R}\right)^{2\alpha}$$

for each $y \in B_{R/2}(x_0)$ and each $\rho \in (0, \tfrac{R}{2}]$. Using the Campanato Lemma, we then have the required result. □

1.4 Harmonic Functions

Recall that a real function u on a domain $\Omega \subset \mathbb{R}^n$ is said to be harmonic if it is C^2 and if $\Delta u = 0$, where $\Delta u := \sum_{j=1}^n D_j D_j u$, with $D_j = \frac{\partial}{\partial x^j}$. Thus we can write $\Delta u = \text{div}(Du)$, where $Du = (D_1 u, \ldots, D_n u)$ as usual denotes the gradient of u, and $\text{div}\,\Phi$ means $\sum_j D_j \Phi^j$ for any smooth vector function $\Phi = (\Phi^1, \ldots, \Phi^n)$.

If we choose a ball $B_R(x_0)$ with closure contained in Ω, and if we integrate the identity $\text{div}\,Du = 0$ over $B_R(x_0)$ and apply the divergence theorem, then we can

1.4. Harmonic Functions

check rather easily (see e.g. [GT83] for the details) that harmonic functions have the mean-value property

(i) $$u(x_0) = \frac{1}{\omega_n R^n} \int_{B_R(x_0)} u$$

for any such ball $B_R(x_0)$. Multiplying (i) by R^n and differentiating with respect to R we get the second version of the mean value property

(i') $$u(x_0) = \frac{1}{\sigma_n R^{n-1}} \int_{S_R(x_0)} u,$$

where $\sigma_n \, (= n\omega_n)$ denotes the measure of the $(n-1)$-dimensional unit sphere $S^{n-1} = \partial B_1(0)$ and where $S_R(x_0) = \partial B_R(x_0)$. These properties are quite fundamental; for example using (i'), one can easily obtain estimates for the partial derivatives of a harmonic function in terms of its L^1 norm:

Lemma 1 *If u is a harmonic function on a domain $\Omega \subset \mathbb{R}^n$ and $\widetilde{\Omega} \subset\subset \Omega$ then $u \in C^\infty$ and there holds for every multi-index α*

(ii) $$\sup_{\widetilde{\Omega}} |D^\alpha u| \leq C \|u\|_{L^1(\Omega)},$$

where the constant C only depends on α and $\mathrm{dist}(\widetilde{\Omega}, \partial\Omega)$.

Proof: Let $R_0 = \mathrm{dist}(\widetilde{\Omega}, \partial\Omega)$ and $\varphi^{(R_0)}$ the mollifier of Section 1.2 having the additional property that $\varphi(x) = \varphi(|x|)$. Then, by multiplying each side of (i') by $\varphi^{(R_0)}(R)$ and integrating each side with respect to R from 0 to R_0, we get for every fixed $y \in \widetilde{\Omega}$ that $u(y) = \int_{\mathbb{R}^n} \varphi^{(R_0)}(x-y) u(x)$, and hence that

$$|D^\alpha u(y)| \leq \int_{B_{R_0}(y)} |D^\alpha_y \varphi^{(R_0)}(x-y)| \, |u(x)| \leq \sup |D^\alpha \varphi^{(R_0)}(x)| \int_\Omega |u|.$$

This completes the proof, because $\sup |D^\alpha \varphi^{(R_0)}(x)| = C_{n,\alpha} R_0^{-|\alpha|}$. □

Remark: In the special case $\Omega = B_R(x_0)$ and $\widetilde{\Omega} = B_{\theta R}(x_0)$ (with $\theta \in (0,1)$ given) one can check that (ii) takes the form

(ii') $$\sup_{B_{\theta R}(x_0)} R^j |D^\alpha u| \leq C_{n,j,\theta} R^{-n} \int_{B_R(x_0)} |u|$$

with $j = |\alpha|$. With a bit of extra effort one can show (by induction on j) that $C_{n,j,\theta} \leq C_{n,\theta}^{j+1} j!$ for a suitable constant $C_{n,\theta}$, and from this (by using Taylor polynomial approximation for u) it follows that in fact u is real-analytic. Thus harmonic functions are automatically real-analytic.

1.5 Weakly Harmonic Functions

Definition 1 *If $u \in W^{1,2}(\Omega)$, where Ω is an open set in $\mathbb{R}^n$, we say that u is weakly harmonic on Ω if*

(i) $$\int_\Omega Du \cdot D\varphi = 0 \quad \forall \varphi \in C_c^\infty(\Omega).$$

Notice this formally generalizes the notion of smooth harmonic, because if u is smooth harmonic then we could integrate by parts in the identity $\int_\Omega \Delta u\, \varphi = 0$, thus establishing (i), and, conversely, if $u \in C^2(\Omega)$, (i) evidently implies $\Delta u = 0$ by virtue of the arbitrariness of φ.

In fact H. Weyl proved that the two notions weakly harmonic and classical harmonic are the same:

Lemma 1 (H. Weyl) *Suppose u is weakly harmonic. Then the L^2 class of u has a C^∞ representative which is harmonic.*

Proof: The key point is to note that if $u \in W^{1,2}(\Omega)$ is weakly harmonic then (with the notation of Section 1.2) u_σ is smooth harmonic on Ω_σ for each $\sigma > 0$. Let us check this claim first. Notice that by differentiation under the integral in the definition of u_σ, we have $\Delta u_\sigma(x) = \int_\Omega u(y) \Delta_x \varphi^{(\sigma)}(x-y)\, dy$. Now by the chain rule $\Delta_x \varphi^{(\sigma)}(x-y) = (-1)^2 \Delta_y \varphi^{(\sigma)}(x-y) = \Delta_y \varphi^{(\sigma)}(x-y)$, and since $\varphi(x-y)$ (as a function of y for x fixed in Ω_σ) is $C_c^\infty(\Omega)$, we use the definition of the weak derivative $D_j u$ to obtain $\int_\Omega u(y) \Delta_x \varphi^{(\sigma)}(x-y)\, dy = -\sum_{j=1}^n \int_\Omega D_{y^j} u(y) D_{y^j}[\varphi(x-y)] = 0$ by definition of weakly harmonic. Thus we have $\Delta u_\sigma = 0$ on Ω_σ as required. Now the rest of the proof follows easily by letting $\sigma \downarrow 0$ and using the Arzela-Ascoli Lemma and the bounds 1.4(ii) with u_σ in place of u. □

1.6 Harmonic Approximation Lemma

The following harmonic approximation (or "blow up") lemma will be of fundamental importance:

Lemma 1 (Harmonic Approximation Lemma) *Let $B = B_1(0)$, the open ball of radius 1 and center 0 in $\mathbb{R}^n$. For each $\varepsilon > 0$ there is $\delta = \delta(n, \varepsilon) > 0$ such that if $f \in W^{1,2}(B)$, $\int_B |Df|^2 \leq 1$ and*

$$\left| \int_B Df \cdot D\varphi \right| \leq \delta \sup_B |D\varphi|, \quad \forall \varphi \in C_c^\infty(B),$$

then there is a harmonic function u on B such that $\int_B |Du|^2 \leq 1$ and

$$\int_B |f - u|^2 \leq \varepsilon^2.$$

Proof: If this fails for some $\varepsilon > 0$, then there is a sequence $\{f_k\} \subset W^{1,2}(B)$, $\int_B |Df_k|^2 \leq 1$,

(1) $$\left| \int_B Df_k \cdot D\varphi \right| \leq k^{-1} \sup_B |D\varphi|$$

for each $\varphi \in C_c^\infty(B)$, and such that

(2) $$\int_B |f_k - u|^2 > \varepsilon^2$$

for every harmonic u on B with $\int_B |Du|^2 \leq 1$.

Notice that since the same holds with $\tilde{f}_k = f_k - \lambda_k$ for any choice of constants λ_k, we can assume without loss of generality that $\int_B f_k = 0$ for each k. But then by using the Poincaré inequality we conclude

$$\limsup_{k \to \infty} \int_B (|f_k|^2 + |Df_k|^2) < \infty,$$

and hence by the Rellich Compactness Theorem (Lemma 1 of Section 1.3) we have a subsequence $f_{k'}$ and an $f \in W^{1,2}(B)$ such that

(3) $$\lim \int_B |f - f_{k'}|^2 = 0$$

and $Df_{k'} \rightharpoonup Df$ weakly in $L^2(B)$. But using this weak convergence in (1) we deduce that $\int_B Df \cdot D\varphi = 0$ for each $\varphi \in C_c^\infty(B)$, so that f is weakly harmonic on B, and Weyl's Lemma guarantees that f is smooth harmonic on B, and hence (since $\int_B |Df|^2 \leq \liminf \int_B |Df_{k'}|^2 \leq 1$) we see that (3) contradicts (2). $\square$

1.7 Elliptic regularity

We here establish the regularity theory and a-priori estimates for the Poisson equation $\Delta u = f$ needed in our later discussion. For more general results (the full Schauder theory), we refer to [GT83].

In this section, and subsequently, $|u|_{k;\Omega}$ denotes the usual norm of $u \in C^k(\overline{\Omega})$, i.e. $|u|_{k;\Omega} = \sum_{|\beta| \leq k} |D^\beta u|_{0;\Omega}$, and $[D^k u]_{\alpha;\Omega} = \sum_{|\beta|=k} [D^\beta u]_{\alpha;\Omega}$.

The following lemma is fundamental:

Lemma 1 *Let $u : \mathbb{R}^n \to \mathbb{R}$ satisfy $[D^2 u]_{\alpha;\mathbb{R}^n} < \infty$. Then there is a constant C_n, depending only on n, such that*

$$[D^2 u]_{\alpha;\mathbb{R}^n} \leq C_n [\Delta u]_{\alpha;\mathbb{R}^n}.$$

Proof: We give a new proof for this fact which is based upon a scaling argument. Assume the assertion is false, i.e. there exist u_k and f_k with $f_k = \Delta u_k$ and $[D^2 u_k]_{\alpha;\mathbb{R}^n} < \infty$ but $[f_k]_{\alpha;\mathbb{R}^n} < \frac{1}{k}[D^2 u_k]_{\alpha;\mathbb{R}^n}$. By definition of $[D^2 u_k]_{\alpha;\mathbb{R}^n}$, for each k we can choose two distinct points $y_k, z_k \in \mathbb{R}^n$ satisfying

$$\frac{|D^2 u_k(y_k) - D^2 u_k(z_k)|}{|y_k - z_k|^\alpha} \geq \frac{1}{2n^2}[D^2 u_k]_{\alpha;\mathbb{R}^n}.$$

Let $\sigma_k := |y_k - z_k|$ and $\lambda_k := [D^2 u_k]_{\alpha;\mathbb{R}^n}$. Since $\lambda_k > 0$ the following scaling is possible:

$$\tilde{u}_k(x) := \lambda_k^{-1} \sigma_k^{-2-\alpha} u_k(y_k + \sigma_k x)$$
$$\tilde{f}_k(x) := \lambda_k^{-1} \sigma_k^{-\alpha} f_k(y_k + \sigma_k x).$$

The scaled functions satisfy $\Delta \tilde{u}_k = \tilde{f}_k$ and $[D^2 \tilde{u}_k]_{\alpha;\mathbb{R}^n} = 1$. Now we investigate the deviation $\hat{u}_k$ of $\tilde{u}_k$ from its 2-jet at the origin: $\hat{u}_k(x) := \tilde{u}_k(x) - \sum_{|\alpha| \leq 2} \frac{x^\alpha}{\alpha!} D^\alpha \tilde{u}_k(0)$. There holds $\Delta \tilde{u}_k = \tilde{f}_k - \tilde{f}_k(0)$ and of course still

(1) $$[D^2 \hat{u}_k]_{\alpha;\mathbb{R}^n} = 1.$$

Furthermore we trivially have

(2) $$D^\alpha \hat{u}_k(0) = 0 \quad \text{for } |\alpha| \leq 2.$$

Considering the vectors $\zeta_k := \frac{z_k - y_k}{|z_k - y_k|} \in S^{n-1}$ we find due to the special choice of y_k and z_k that

(3) $$|D^2 \hat{u}_k(\zeta_k)| \geq \frac{1}{2n^2}.$$

(1) and (2) in particular tell us that $\{\hat{u}_k\}$ has second derivatives forming an equicontinuous family and has all derivatives of order up to and including 2 equal to zero at $x = 0$. Thus from an appropriate version of the Arzela-Ascoli Lemma and the compactness of S^{n-1} respectively we infer that there exists a function $v \in C^{2,\alpha}(\mathbb{R}^n)$, a vector $\zeta \in S^{n-1}$ and a subsequence $\{k'\}$ such that $\hat{u}_{k'} \to v$ locally in C^2 and $\zeta_k \to \zeta$ in $\mathbb{R}^n$. By the lower semicontinuity of the seminorm $[\cdot]_{\alpha;\mathbb{R}^n}$ and (1) we get $[D^2 v]_{\alpha;\mathbb{R}^n} \leq 1$, and from (2) we get $D^2 v(0) = 0$, while $D^2 v(\zeta) \neq 0$ from (3). Since $[\tilde{f}_k]_{\alpha;\mathbb{R}^n} \leq \frac{1}{k}$ we get $\sup_{B_R(0)} |\tilde{f}_k - \tilde{f}_k(0)| \leq \frac{1}{k} R^\alpha \to 0$ for every fixed R and thus $\Delta v = 0$ which then implies $\Delta D^2 v = 0$. Using the estimate (ii') of Section 1.4 for the harmonic function $D^2 v$ we obtain

$$\sup_{B_{R/2}(0)} |D^3 v| \leq \frac{C_n}{R} \sup_{B_R(0)} |D^2 v| = \frac{C_n}{R} \sup_{B_R(0)} |D^2 v - D^2 v(0)|$$

$$\leq \frac{C_n}{R} [D^2 v]_{\alpha;\mathbb{R}^n} R^\alpha \leq C_n R^{\alpha-1} \to 0 \quad \text{as } R \to \infty.$$

Thus $D^3 v \equiv 0 \in \mathbb{R}^n$, i.e. v is a polynomial of degree ≤ 2. But $D^2 v(0) = 0$ and hence we conclude $D^2 v \equiv 0$. This contradicts (3). □

We note that there is also a $C^{1,\alpha}$ version of Lemma 1 which can be proved very easily by a similar scaling argument:

1.8. A Technical Regularity Lemma

Lemma 2 *Suppose* $[Du]_{\alpha;\mathbb{R}^n} < \infty$. *Then*

$$[Du]_{\alpha;\mathbb{R}^n} \leq C[\Delta u]^*_{\alpha;\mathbb{R}^n}$$

where $[g]^*_{\alpha;\mathbb{R}^n} = \inf \sum_j [f_j]_{\alpha;\mathbb{R}^n}$ *with the* inf *taken over all collections* $f_1, \ldots, f_n$ *such that* $g = \sum_{j=1}^n D_j f_j$ *(weakly) on* $\mathbb{R}^n$.

We shall need to use the following standard differentiability theory for solutions of $\Delta u = f$. A proof can be based on Lemmas 1 and 2 together with mollification. (For complete details and more general results, we refer to e.g. [GT83, Chapters 6, 8].)

Lemma 3 (Differentiability Theorem) *Suppose* $u \in W^{1,2}(B_R(x_0))$ *is a weak solution of* $\Delta u = f$ *in* $B_R(x_0)$. *Then*

(i) f *bounded on* $B_R(x_0)$ $\Rightarrow$ $u \in C^{1,\alpha}(B_R(x_0))$ $\forall \alpha < 1$,

(ii) $\forall \alpha \in (0,1), k \in \{0, 1, \ldots\} : f \in C^{k,\alpha}(B_R(x_0))$ $\Rightarrow$ $u \in C^{k+2,\alpha}(B_R(x_0))$;

in each case there are corresponding estimates, so that we have the additional conclusions

$$R^{1+\alpha}[Du]_{\alpha;B_{\theta R}(x_0)} \leq C(|u|_{0;B_R(x_0)} + R^2|f|_{0;B_R(x_0)}),$$

with $C = C(n, \theta, \alpha)$, *in case (i), and*

$$\sum_{j=1}^{k+2} R^j |D^j u|_{0;B_{\theta R}} + R^{k+2+\alpha}[D^{k+2} u]_{\alpha;B_{\theta R}(x_0)} \leq$$

$$\leq C(|u|_{0;B_R(x_0)} + R^2|f|_{0;B_R(x_0)} + R^{2+k+\alpha}[D^k f]_{\alpha;B_R(x_0)}),$$

$C = C(n, \theta, \alpha, k)$, *in case (ii)*.

1.8 A Technical Regularity Lemma

Here we shall prove the following general regularity lemma, on which our later proof (in Section 2.3) of the ε-regularity theorem for energy minimizing maps will be based.

Lemma 1 (Technical Lemma) *Suppose* $\alpha \in (0,1)$ *and* $\beta \geq 1$ *are given. There exists* $\delta_0 = \delta_0(n, \alpha, \beta) > 0$ *such that the following holds: If* $u = (u^1, \ldots, u^p) \in W^{1,2}(B_R(x_0); \mathbb{R}^p)$ *satisfies the equation*

$$\Delta u = F \quad \text{weakly in } B_R(x_0),$$

where $F \in L^1(B_R(x_0))$ *with*

(i) $\qquad |F(x)| \leq \beta |Du(x)|^2$ *a.e.* $x \in B_R(x_0)$, *if u satisfies the "reverse Poincaré" inequality*

(ii) $$\left(\frac{\rho}{2}\right)^{2-n} \int_{B_{\rho/2}(y)} |Du|^2 \leq \beta \rho^{-n} \int_{B_\rho(y)} |u - \lambda_{y,\rho}|^2$$

whenever $B_\rho(y) \subset B_R(x_0)$, and if

(iii) $$R^{-n} \int_{B_R(x_0)} |u - \lambda_{x_0,R}|^2 \leq \delta_0^2,$$

then $u \in C^{1,\alpha}(\overline{B}_{R/4}(x_0))$ with

$$|u|_{1;B_{R/4}(x_0)} + [Du]_{\alpha;B_{R/4}(x_0)} \leq C \left(R^{-n} \int_{B_R(x_0)} |u - \lambda_{x_0,R}|^2 \right)^{1/2},$$

where C is a constant depending only on n, α and β.

Remark: Notice that $\Delta u = F$ weakly means that

$$\int_{B_R(x_0)} \sum_{j=1}^n D_j u \cdot D_j \varphi = -\int_{B_R(x_0)} \varphi \cdot F, \quad \forall \varphi \in C_c^\infty(B_R(x_0)).$$

Before we begin the proof of this lemma, we need to mention the appropriately scaled version of the harmonic approximation lemma discussed in Section 1.6.

Lemma 2 (Rescaled Version of the Harmonic Approximation Lemma)
For any given $\varepsilon > 0$ there is $\delta = \delta(n, \varepsilon) > 0$ such that if $f \in W^{1,2}(B_\rho(y))$, if $\rho^{2-n} \int_{B_\rho(y)} |Df|^2 \leq 1$, and if $|\rho^{2-n} \int_{B_\rho(y)} Df \cdot D\varphi| \leq \delta\rho \sup_{B_\rho(y)} |D\varphi|$ for every $\varphi \in C_c^\infty(B_\rho(y))$, then there is a harmonic function u on $B_\rho(y)$ with $\rho^{2-n} \int_{B_\rho(y)} |Du|^2 \leq 1$ and $\rho^{-n} \int_{B_\rho(y)} |u - \lambda_{y,\rho}|^2 \leq \varepsilon^2$.

Notice that this easily follows from the unscaled version as in Lemma 1 of Section 1.6; in fact one just checks that Lemma 1 of Section 1.6 applies to the rescaled function $f_\rho(x) \equiv f(y + \rho(x - y))$.

Proof of the Technical Lemma: As in the above remark we have

(1) $$\int_{B_R(x_0)} Du \cdot D\varphi = -\int_{B_R(x_0)} \varphi \cdot F, \quad \forall \varphi \in C_c^\infty(B_R(x_0)).$$

Let $B_\rho(y) \subset B_R(x_0)$ be an arbitrary ball. By using identity (1) with $\varphi \in C_c^\infty(B_{\rho/2}(y))$, and using also the hypotheses (i), (ii), we have

$$\left| \left(\frac{\rho}{2}\right)^{2-n} \int_{B_{\rho/2}(y)} Du \cdot D\varphi \right| \leq \beta^2 \rho^{-n} \sup_{B_{\rho/2}(y)} |\varphi| \int_{B_\rho(y)} |u - \lambda_{y,\rho}|^2.$$

1.8. A Technical Regularity Lemma

Since $\sup_{B_{\rho/2}(y)} |\varphi| \leq \rho \sup_{B_{\rho/2}(y)} |D\varphi|$ (by 1-dimensional calculus along line segments in $B_{\rho/2}(y)$), we thus have

$$\left| \left(\frac{\rho}{2}\right)^{2-n} \int_{B_{\rho/2}(y)} Dv \cdot D\varphi \right| \leq \left(\rho^{-n} \int_{B_\rho(y)} |u - \lambda_{y,\rho}|^2 \right)^{1/2} \rho \sup_{B_{\rho/2}(y)} |D\varphi| \quad (2)$$

where $v = \ell^{-1} u$, with $\ell = \beta \left(\rho^{-n} \int_{B_\rho(y)} |u - \lambda_{y,\rho}|^2 \right)^{1/2}$.

Also, we have trivially by (ii) and the definition of v that

$$\left(\frac{\rho}{2}\right)^{2-n} \int_{B_{\rho/2}(y)} |Dv|^2 \leq 1. \quad (3)$$

Let $\varepsilon > 0$ be for the moment arbitrary. By (2) and (3), we can apply the harmonic approximation lemma (Lemma 2 above) in order to conclude that there is a harmonic function w on $B_{\rho/2}(y)$ such that

$$\left(\frac{\rho}{2}\right)^{2-n} \int_{B_{\rho/2}(y)} |Dw|^2 \leq 1 \text{ and } \left(\frac{\rho}{2}\right)^{-n} \int_{B_{\rho/2}(y)} |v - w|^2 \leq \varepsilon^2, \quad (4)$$

assuming that $\rho^{-n} \int_{B_\rho(y)} |u - \lambda_{y,\rho}|^2 \leq \delta^2$, where $\delta = \delta(n, \varepsilon)$ is as in the harmonic approximation lemma. Now take $\theta \in (0, \frac{1}{4}]$ and note that by the squared triangle inequality

$$(\theta \rho)^{-n} \int_{B_{\theta\rho}(y)} |v - w(y)|^2 \leq 2(\theta\rho)^{-n} \int_{B_{\theta\rho}(y)} \left(|v - w|^2 + |w - w(y)|^2 \right). \quad (5)$$

Now using 1-dimensional calculus along line segments with end-point at y together with the estimate 1.4(ii') with $j = 0$ (applied to $D_i w$), we have

$$\sup_{B_{\theta\rho}(y)} |w - w(y)|^2 \leq (\theta\rho \sup_{B_{\theta\rho}(y)} |Dw|)^2 \leq C\theta^2 \rho^{2-n} \int_{B_{\rho/2}(y)} |Dw|^2.$$

Using this together with (4) in (5), we conclude that

$$(\theta\rho)^{-n} \int_{B_{\theta\rho}(y)} |v - w(y)|^2 \leq \theta^{-n} \varepsilon^2 + C\theta^2,$$

where C depends only on n. Writing $v = \ell^{-1} u$, we have

$$(\theta\rho)^{-n} \int_{B_{\theta\rho}(y)} |u - \lambda|^2 \leq \beta^2 (\theta^{-n} \varepsilon^2 + C\theta^2) \rho^{-n} \int_{B_\rho(y)} |u - \lambda|^2,$$

where $\lambda = \ell w(y)$ is a fixed vector in $\mathbb{R}^p$. Now we choose θ and ε: first select $\theta \in (0, \frac{1}{4}]$ so that $C\beta^2 \theta^2 \leq \frac{1}{2}\theta^{2\alpha}$; notice that such θ can be chosen to depend only on n, α, β. Having so chosen θ, now choose $\varepsilon > 0$ such that $\beta^2 \theta^{-n} \varepsilon^2 < \frac{1}{2}\theta^{2\alpha}$. Therefore

$$(\theta\rho)^{-n} \int_{B_{\theta\rho}(y)} |u - \lambda_{y,\theta\rho}|^2 \leq \theta^{2\alpha} \rho^{-n} \int_{B_\rho(y)} |u - \lambda_{y,\rho}|^2.$$

Thus, to summarize, we have shown that if $B_\rho(y) \subset B_R(x_0)$, and if $\rho^{-n} \int_{B_\rho(y)} |u - \lambda_{y,\rho}|^2 \leq \delta_0^2$, with $\delta_0 = \delta_0(n, \alpha, \beta)$ sufficiently small, then

$$(6) \qquad (\theta\rho)^{-n} \int_{B_{\theta\rho}(y)} |u - \lambda_{y,\theta\rho}|^2 \leq \theta^{2\alpha} \rho^{-n} \int_{B_\rho(y)} |u - \lambda_{y,\rho}|^2,$$

where $\theta = \theta(n, \alpha, \beta) \in (0, \tfrac{1}{4}]$. From now on we assume δ_0 has been so chosen.

Next, let $I_0 = R^{-n} \int_{B_R(x_0)} |u - \lambda_{x_0,R}|^2$ and note that if $I_0 \leq 2^{-n} \delta_0^2$, then, with $\rho = R/2$, we have $B_\rho(y) \subset B_R(x_0)$ and

$$\rho^{-n} \int_{B_\rho(y)} |u - \lambda_{x_0,R}|^2 \leq 2^n I_0 < \delta_0^2 \qquad \forall y \in B_{R/2}(x_0).$$

Thus, if $I_0 < 2^{-n} \delta_0^2$, then

$$(7) \qquad \rho^{-n} \int_{B_\rho(y)} |u - \lambda_{y,\rho}|^2 \leq \delta_0^2, \quad \forall y \in B_{R/2}(x_0),$$

where $\rho = R/2$. But notice that this in particular means that the "starting hypothesis" $\rho^{-n} \int_{B_\rho(y)} |u - \lambda_{y,\rho}|^2 < \delta_0^2$ is satisfied with $\theta\rho$ in place of ρ, and hence (6) holds with $\theta\rho$ in place of ρ ($\rho = R/2$ still). Continuing inductively, we deduce that

$$(8) \qquad \left(\theta^j \frac{R}{2}\right)^{-n} \int_{B_{\theta^j R/2}(y)} |u - \lambda_{y,\theta^j R/2}|^2 \leq \theta^{2j\alpha} \left(\frac{R}{2}\right)^{-n} \int_{B_{R/2}(y)} |u - \lambda_{y,R/2}|^2 \leq 2^n I_0$$

for each $j = 0, 1, 2, \ldots$, provided only that the inequality $2^n I_0 < \delta_0^2$ does hold. Now on the other hand if $\sigma \in (0, R/2]$, there is a unique $j \in \{0, 1, \ldots\}$ such that $\theta^{j+1} R/2 < \sigma \leq \theta^j R/2$, and it is then easy to check that (8) actually implies

$$\sigma^{-n} \int_{B_\sigma(y)} |u - \lambda_{y,\sigma}|^2 \leq 2^{n+2\alpha} \theta^{-n-2\alpha} \left(\frac{\sigma}{R}\right)^{2\alpha} I_0 \qquad \forall y \in B_{R/2}(x_0), \ \sigma \in (0, R/2],$$

provided $2^n I_0 < \delta_0^2$. Then by virtue of the Campanato Lemma (Lemma 1 of Section 1.1) we have $u \in C^{0,\alpha}(\overline{B}_{R/2}(x_0))$ and

$$(9) \qquad [u]_{\alpha; B_{R/2}(x_0)} \leq C \left(R^{-n} \int_{B_R(x_0)} |u - \lambda_{x_0,R}|^2\right)^{1/2}$$

Next we want to show that $u \in C^{1,\alpha}(\overline{B}_{R/4}(x_0))$ as claimed in the statement of the lemma. Since we may change scale, it suffices to prove this in case $R = 1$; so we assume here that the hypotheses hold with $R = 1$. First let $B_\rho(y) \subset B_{1/2}(x_0)$ be arbitrary. Since $u \in W^{1,2}(B_\rho(y)) \cap C^0(\overline{B}_\rho(y))$, it is standard (see e.g. [GT83]) that there is a $v \in C^2(B_\rho(y); \mathbb{R}^p) \cap C^0(\overline{B}_\rho(y); \mathbb{R}^p) \cap W^{1,2}(B_\rho(y); \mathbb{R}^p)$ which is harmonic on $B_\rho(x_0)$ and which agrees with u on $\partial B_\rho(y)$. Of course then v satisfies the weak form of Laplace's equation on $B_\rho(y)$; that is,

$$(10) \qquad \int_{B_\rho(y)} \sum_{j=1}^n D_j v \cdot D_j \varphi = 0, \quad \forall \varphi \in C_c^\infty(B_\rho(y); \mathbb{R}^p).$$

1.8. A Technical Regularity Lemma

Taking the difference of the equations (1) and (10), we thus get that

$$\int_{B_\rho(y)} \sum_{j=1}^n D_j(u-v) \cdot D_j\varphi = \int_{B_\rho(y)} \varphi \cdot F, \quad \forall \varphi \in C_c^\infty(B_\rho(y); \mathbb{R}^p).$$

Now since u and v agree on the boundary it is easy to check that this is also valid with the choice $\varphi = v - u$; here we use the general fact that if a $C^0(\overline{B}_\rho(y)) \cap W^{1,2}(B_\rho(y))$ function is zero on $\partial B_\rho(y)$, then it is the limit in the $W^{1,2}(B_\rho(y))$ norm of a sequence of $C_c^\infty(B_\rho(y))$ functions. Thus we obtain

(11)
$$\int_{B_\rho(y)} |D(u-v)|^2 = \int_{B_\rho(y)} (u-v) \cdot F \le$$

$$\le \beta \sup_{B_\rho(y)} |u-v| \int_{B_\rho(y)} |Du|^2 \le C\rho^\alpha \int_{B_\rho(y)} |Du|^2,$$

where we used the fact that $\sup_{B_\rho(y)} |u-u(y_0)| \le C\rho^\alpha$ (from the Hölder continuity of u proved above) for any $y_0 \in \overline{B}_\rho(y)$ and that, for $y_0 \in \partial B_\rho(y)$, $\sup_{B_\rho(y)} |v-u(y_0)| \equiv \sup_{\partial B_\rho(y)} |v-v(y_0)| \le n \sup |u-u(y_0)| \le C\rho^\alpha$ by applying the maximum principle to each component v^j of $v = (v^1, \ldots, v^p)$. By using the reverse Hölder inequality and the Hölder estimate (9), we have

$$\int_{B_\rho(y)} |Du|^2 \le C\rho^{-2} \int_{B_{2\rho}(y)} |u-\lambda_{y,2\rho}|^2 \le C\rho^{-2+2\alpha} \int_{B_1(x_0)} |u-\lambda_{x_0,R}|^2,$$

and hence (11) gives

(12)
$$\rho^{-n} \int_{B_\rho(y)} |D(u-v)|^2 \le C\ell^2 \rho^{3\alpha-2}, \quad \ell = \left(\int_{B_1(x_0)} |u-\lambda_{x_0,1}|^2 \right)^{1/2},$$

for such ρ. Now let us agree that α was chosen in the first place so that $3\alpha > 1$, and that hence $3\alpha = 1 + 2\gamma$ for some $\gamma > 0$. Thus we have

$$\rho^{-n} \int_{B_\rho(y)} |D(u-v)|^2 \le C\ell^2 \rho^{3\alpha-2}, \quad \text{for } B_\rho(y) \subset B_1(x_0).$$

Therefore we get for any $\sigma \le \rho$

$$\sigma^{-n} \int_{B_\sigma(y)} |Du - Dv(y)|^2 \le$$

$$\le 2\sigma^{-n} \int_{B_\sigma(y)} |Du - Dv|^2 + 2\sigma^{-n} \int_{B_\sigma(y)} |Dv - Dv(y)|^2$$

$$\le 2\sigma^{-n} \int_{B_\rho(y)} |Du - Dv|^2 + 2\sigma^{-n} \int_{B_\sigma(y)} |Dv - Dv(y)|^2$$

$$\le C\ell^2 \left(\frac{\rho}{\sigma}\right)^n \rho^{3\alpha-2} + C\sigma^2 \sup_{B_\sigma(y)} |D^2 v|^2.$$

By using the inequality 1.4(ii') with $R = \rho/2$ and the harmonic functions $D_j v - \Lambda_j$ in place of u (Λ_j any constants) we have

$$\sigma^2 \sup_{B_\sigma(y)} |D^2 v|^2 \leq \sigma^2 \sup_{B_{\rho/2}(y)} |D^2 v|^2 + C\sigma^2 \rho^{-2} \rho^{-n} \int_{B_{3\rho/4}(y)} |Dv - \Lambda|^2,$$

with any constant $\Lambda \in \mathbb{R}^p$. Taking $\Lambda = \Lambda_{y,\rho} = |B_\rho(y)|^{-1} \int_{B_\rho(y)} Du$, and using (12) again, we get

$$\sigma^2 \sup_{B_\sigma(y)} |D^2 v|^2 \leq$$

$$\leq C\sigma^2 \rho^{-2} \rho^{-n} \int_{B_{3\rho/4}(y)} |Dv - \Lambda_{y,\rho}|^2$$

$$\leq C\sigma^2 \rho^{-2} \rho^{-n} \int_{B_{3\rho/4}(y)} |Dv - Du|^2 + C\sigma^2 \rho^{-2} \rho^{-n} \int_{B_{3\rho/4}(y)} |Du - \Lambda_{y,\rho}|^2$$

$$\leq C\ell^2 \left(\frac{\sigma}{\rho}\right)^2 \rho^{3\alpha - 2} + C\left(\frac{\sigma}{\rho}\right)^2 \rho^{-n} \int_{B_\rho(y)} |Du|^2.$$

Combining all the above estimates and setting $I_\rho := \rho^{-n} \int_{B_\rho(y)} |Du - \Lambda_{\rho,y}|^2$, we conclude

$$I_\sigma \leq C\ell^2 \left(\frac{\rho}{\sigma}\right)^n \rho^{3\alpha - 2} + C\ell^2 \left(\frac{\sigma}{\rho}\right)^2 \rho^{3\alpha - 2} + C\left(\frac{\sigma}{\rho}\right)^2 \rho^{-n} \int_{B_\rho(y)} |Du|^2.$$

But by the reverse Poincaré inequality (hypothesis (ii) of the technical lemma) we get

$$\rho^{-n} \int_{B_\rho(y)} |Du|^2 \leq C\rho^{-n} \int_{B_{2\rho}(y)} |u - \lambda_{y,2\rho}|^2 \leq C\ell^2 \rho^{-2 + 2\alpha}$$

by the Hölder estimate from the first part of the proof. So using this on the right of the previous inequality we get

$$I_\sigma \leq C\ell^2 \left(\frac{\rho}{\sigma}\right)^n \rho^{3\alpha - 2} + C\ell^2 \left(\frac{\sigma}{\rho}\right)^2 \rho^{-2 + 2\alpha}.$$

This holds for all $\sigma \leq \rho < 1/8$, $y \in B_{3/8}(x_0)$. Choosing $\sigma = \rho^\kappa$, $\kappa := 1 + \frac{\alpha}{n+2}$, we can rewrite this in the form

(13) $$\sigma^{-n} \int_{B_\sigma(y)} |Du - \Lambda_{y,\sigma}|^2 \leq C\ell^2 \sigma^{2\gamma}, \quad \sigma \leq (1/8)^\kappa, \; y \in B_{3/8}(x_0),$$

where $2\gamma = (\frac{2\alpha}{n+2} + 2\alpha - 2)/\kappa > 0$ if we choose α close to 1. From this we get $[Du]_{\gamma; B_{3/8}(x_0)} \leq C\ell$ by virtue of Campanato's Lemma (Lemma 1 of Section 4). In particular, since $\int_{B_1(x_0)} |Du|^2 \leq \Lambda$, this establishes $\sup_{B_{3/8}(x_0)} |Du| \leq C\ell$, and hence the equation has the form $\Delta u = f$ with f bounded by $C\ell$ on $B_{3/8}(x_0)$. Then the general assertion follows by using Lemma 3 of Section 1.7 with $u - u(x_0)$ in place of u, since $\sup_{B_{3/8}(x_0)} |u - u(x_0)| \leq C[u]_{\alpha; B_{3/8}(x_0)} \leq C\ell$ by (9). $\square$

Later (in Section 2.3) we shall apply this technical lemma to study energy minimizing maps.

Chapter 2

Regularity Theory for Harmonic Maps

2.1 Definition of Energy Minimizing Maps

Suppose that Ω is an open subset of $\mathbb{R}^n$, $n \geq 2$, and that N is a smooth compact Riemannian manifold of dimension $m \geq 2$ which is isometrically embedded in some Euclidean space $\mathbb{R}^p$. We look at maps u of Ω into N; such a map will always be thought of as a map $u = (u^1, \ldots, u^p) : \Omega \to \mathbb{R}^p$ with the additional property that $u(\Omega) \subset N$.

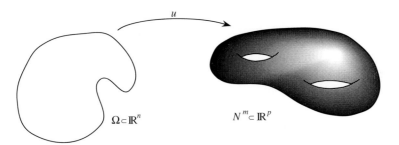

We do not assume here that u is smooth—in fact we make only the minimal assumption necessary to ensure that the energy of u is well-defined. This gives rise to the following definition:

Definition 1 (i) *For Ω and N as specified above, the Sobolev space $W^{1,2}_{\text{loc}}(\Omega; N)$ is defined as the set of functions $u \in W^{1,2}_{\text{loc}}(\Omega; \mathbb{R}^p)$ with $u(x) \in N$ a.e. $x \in \Omega$.*

(ii) *The energy $\mathcal{E}_{B_\rho(y)}(u)$ for a function $u \in W^{1,2}_{\text{loc}}(\Omega; N)$ in a ball $B_\rho(y) := \{x : |x - y| < \rho\}$ with $\overline{B}_\rho(y) \subset \Omega$ is defined by*

$$\mathcal{E}_{B_\rho(y)}(u) = \int_{B_\rho(y)} |Du|^2.$$

Notice that here Du means the $n \times p$ matrix with entries $D_i u^j (:= \partial u^j/\partial x^i)$, and $|Du|^2 = \sum_{i=1}^{n} \sum_{j=1}^{p} (D_i u^j)^2$. We study maps $u \in W^{1,2}(B_\rho(y); N)$ which minimize energy in Ω in the sense that, for each ball $B_\rho(y) \subset \Omega$,

$$\mathcal{E}_{B_\rho(y)}(u) \leq \mathcal{E}_{B_\rho(y)}(w),$$

for every $w \in W^{1,2}(B_\rho(y); N)$ with $w \equiv u$ in a neighbourhood of $\partial B_\rho(y)$. Such a u will be called an energy minimizing map into N.

Remark: The theory to be developed in the sequel may be applied, after very minor modifications, to the case when Ω is equipped with a general smooth Riemannian metric $\sum_{i,j} g_{ij} dx^i dx^j$ rather than with the standard Euclidean metric; see the further discussion in Section 4.8 below.

2.2 The Variational Equations

Suppose u is energy minimizing as in Section 2.1, suppose $\overline{B}_\rho(y) \subset \Omega$, and suppose that for some $\delta > 0$ we have a 1-parameter family $\{u_s\}_{s \in (-\delta,\delta)}$ of maps of $B_\rho(y)$ into N such that $u_0 = u$, $Du_s \in L^2(\Omega)$, and $u_s \equiv u$ in a neighbourhood of $\partial B_\rho(y)$ for each $s \in (-\delta, \delta)$. Then by definition of minimizing we know $\mathcal{E}_{B_\rho(y)}(u_s)$ takes its minimum at $s = 0$, and hence

(i) $$\left.\frac{d\mathcal{E}_{B_\rho(y)}(u_s)}{ds}\right|_{s=0} = 0$$

whenever the derivative on the left exists. The derivative on the left is called the first variation of $\mathcal{E}_{B_\rho(y)}$ relative to the given family; the family $\{u_s\}$ itself is called an (admissible) variation of u. There are two important kinds of variations of u:

Class 1: Variations of the form

(ii) $$u_s = \Pi \circ (u + s\zeta),$$

where $\zeta = (\zeta^1, \ldots, \zeta^p)$ with each $\zeta^j \in C_c^\infty(B_\rho(y))$ and where Π is the nearest point projection onto N. Notice (see Appendix 2.12.3 below) that nearest point projection onto N is well-defined and smooth in some open subset W containing N, and hence u_s defined in (ii) is an admissible variation for $|s|$ small enough: see Fig. 2.1). Now by applying D_i to the Taylor polynomial expansion of Π we have $D_i u_s = D_i u + s(d\Pi_u(D_i \zeta) + \text{Hess}\,\Pi_u(\zeta, D_i u)) + E$, where $\|E\|_{L^1(B_\rho(y))} \leq Cs^2$ for $|s|$ small. Plugging this expression for $D_i u_s$ into the energy $\mathcal{E}_{B_\rho(y)}(u_s)$ and using the facts about the induced linear map and Hessian of Π given in (iii), (iv), (v) of Theorem 1 of Appendix 2.12.3, we check that for such a variation the equation (i) gives the integral identity

(iii) $$\int_\Omega \sum_{i=1}^{n} (D_i u \cdot D_i \zeta - \zeta \cdot A_u(D_i u, D_i u)) = 0$$

2.2. The Variational Equations

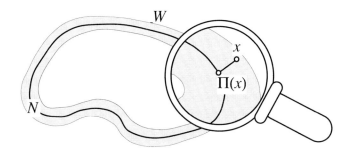

Figure 2.1: *Nearest point projection*

for any ζ as above. Notice that if u is C^2 we can integrate by parts here and use the fact that ζ is an arbitrary C^∞ function in order to deduce the equation

(iv) $$\Delta u + \sum_{i=1}^{n} A_u(D_i u, D_i u) = 0,$$

where Δu means simply $(\Delta u^1, \ldots, \Delta u^p)$. The identity (iii) is called the weak form of the equation (iv); of course if u is not C^2 the equation (iv) makes no sense classically, and *must* be interpreted in the weak sense (iii). It is worth noting that, in case $u \in C^2$, (iii) says simply

(iv') $$(\Delta u)^T = 0$$

at a given point $x \in B_\rho(y)$, where $(\Delta u)^T$ means orthogonal projection of $\Delta u(x)$ onto the tangent space $T_{u(x)}N$ of N at the image point $u(x)$. This follows directly from the general identity (vi) in Theorem 1 of Appendix 2.12.3.

Class 2: Variations of the form

$$u_s(x) = u(x + s\zeta(x)),$$

where $\zeta = (\zeta^1, \ldots, \zeta^n)$ with each $\zeta^j \in C_c^\infty(B_\rho(y))$. Then $D_i u_s(x) = \sum_{j=1}^{n} D_i u(x + s\zeta) + s D_i \zeta^j D_j u(x + s\zeta)$, and hence after making the change of variable $\xi = x + s\zeta$ (which gives a C^∞ diffeomorphism of $B_\rho(y)$ onto itself in case $|s|$ is small enough) in this case (i) implies

(v) $$\int_{B_\rho(y)} \sum_{i,j=1}^{n} \left(|Du|^2 \delta_{ij} - 2 D_i u \cdot D_j u\right) D_i \zeta^j = 0.$$

(Notice that in checking this we need to observe that the Jacobian determinant of the transformation $x \mapsto \xi = x + s\zeta(x)$ satisfies $|\det(\partial x^i / \partial \xi^j)| = |\det(\partial \xi^i / \partial x^j)|^{-1} = 1 - s \operatorname{div} \zeta + O(s^2)$.) The identities (iii), (v) are of great importance in the study of energy minimizing maps. Notice that if $u \in C^2$ we can integrate by parts in (v) in order to deduce that (iii) *implies* (v); it is however false that (iii) implies (v) in case Du is merely in L^2 (and there are simple examples to illustrate this). One calls a map u into N which satisfies (iii) a "weakly harmonic map", while a map which satisfies both (iii) and (v) is usually referred to as a "stationary harmonic map". The above discussion thus proves that energy minimizing implies stationary harmonic. We shall not here discuss weakly harmonic maps, but we do mention that such maps admit far worse singularities than the energy minimizing maps—see e.g. [Riv92]—except in the case $n = 2$ when there are no singularities at all. We show this below in the case of minimizing maps, and refer to recent work of F. Hélein [He91] for the general case of weakly harmonic maps.

2.3 The ε-Regularity Theorem

We can now state the Schoen-Uhlenbeck regularity Theorem:

Theorem 1 (ε-Regularity Theorem) *Let $\Lambda > 0$, $\theta \in (0, 1)$. There exists $\varepsilon = \varepsilon(n, N, \Lambda, \theta) > 0$ such that if $u \in W^{1,2}(\Omega; N)$ is energy minimizing on $B_R(x_0) \subset \Omega$ and if $R^{2-n} \int_{B_R(x_0)} |Du|^2 \leq \Lambda$ and $R^{-n} \int_{B_R(x_0)} |u - \lambda_{x_0,R}|^2 < \varepsilon^2$, then there holds $u \in C^{\infty}(B_{R/4}(x_0))$, and for $j = 1, 2, \ldots$ we have the estimates*

$$R^j \sup_{B_{\theta R}(x_0)} |D^j u| \leq C \Big(R^{-n} \int_{B_R(x_0)} |u - \lambda_{x_0,R}|^2\Big)^{1/2},$$

where C depends only on j, Λ, N, θ, and n.

Remark: It suffices to prove the lemma for some fixed θ (e.g. $\theta = \frac{1}{8}$). To see this, suppose the lemma is proved with $\theta = \frac{1}{8}$, and select $Q = Q(n, \theta)$ and points $y_1, \ldots, y_Q \in B_{\theta R}(x_0)$ such that $B_{\theta R}(x_0) \subset \cup_j^Q B_{(1-\theta)R/8}(y_j)$. Thus we can apply the theorem with $\theta = \frac{1}{8}$, with y_j in place of x_0, and with $(1-\theta)R$ in place of R. Since $B_{(1-\theta)R}(y_j) \subset B_R(x_0)$ for each j, the required bounds on $\sup_{B_{\theta R}(x_0)} |D^j u|$ then follow because $\int_{B_{(1-\theta)R}(y_j)} |u - \lambda_{y_j, (1-\theta)R}|^2 \leq \int_{B_{(1-\theta)R}(y_j)} |u - \lambda_{x_0,R}|^2 \leq \int_{B_R(x_0)} |u - \lambda_{x_0,R}|^2$.

Proof of Theorem 1: In view of the above remark, it will suffice to prove the theorem in the special case $\theta = \frac{1}{8}$. According to the discussion in Section 2.2 above we know that u satisfies $\Delta u + \sum_{j=1}^{n} A_u(D_j u, D_j u) = 0$ weakly in $B_R(x_0)$. But this can be written in the form $\Delta u = F$, where $|F| = |\sum_{j=1}^{n} A_u(D_j u, D_j u)| \leq C|Du|^2$ with $C = n \sup_{y \in N, \tau \in T_y N, |\tau|=1} |A_y(\tau, \tau)|$ depending only on n and N. Thus the technical Lemma 1 in Section 1.8 gives immediately that $u \in C^{1,\alpha}(\overline{B}_{R/4}(x_0))$ for any $\alpha \in (0, 1)$, and

$$(1) \quad |u|_{1; B_{R/4}(x_0)} + [Du]_{\alpha; B_{R/4}(x_0)} \leq C_{n,N,\alpha} \Big(R^{-n} \int_{B_R(x_0)} |u - \lambda_{x_0,R}|^2\Big)^{1/2},$$

*provided u satisfies the reverse Poincaré hypothesis (ii) from the technical lemma. The fact that $u \in C^{\infty}(B_{R/4}(x_0))$ (and the stated estimates) now follows from Lemma 3 of Section 1.7 as follows:

We have

$$\Delta u = \sum_{j=1}^{n} A_u(D_j u, D_j u). \tag{2}$$

Since N is smooth, we know that $A_y(\cdot, \cdot)$ is a smooth function of its arguments, and, since $u \in C^{1,\alpha}(\overline{B}_{R/4}(x_0))$, it then follows that $\sum_{j=1}^{n} A_u(D_j u, D_j u)$ is of class $C^{0,\alpha}(\overline{B}_{R/4}(x_0))$. That is the right side of (2) is in $C^{0,\alpha}(\overline{B}_{R/4}(x_0))$ and hence, by Lemma 3 of Section 1.7, we deduce that $u \in C^{2,\alpha}(B_{R/4}(x_0))$. But then the right side of (2) is of class $C^{1,\alpha}(B_{R/4}(x_0))$ and hence by Lemma 3 again we have $u \in C^{3,\alpha}(B_{R/4}(x_0))$. Continuing inductively we conclude that $u \in C^{\infty}(B_{R/4}(x_0))$ as required. Furthermore at the same time (using the estimates of Lemma 3 of Section 1.7), we can inductively check that

$$R^j \sup_{B_{(\frac{1}{8}+\frac{1}{j})R}(x_0)} |D^j u| \leq C \left(R^{-n} \int_{B_R(x_0)} |u - \lambda_{x_0,R}|^2 \right)^{1/2}$$

for each $j = 1, 2, \ldots$.

Thus the theorem (with $\theta = \frac{1}{8}$) is proved modulo checking that there is a "reverse Hölder inequality" like that in hypothesis (ii) of the technical lemma (Lemma 1 in Section 1.8). We defer the proof of this until Section 2.8, when we shall have more theory at our disposal.

2.4 The Monotonicity Formula

An important consequence of the variational identity (v) of Section 2.2 is the "monotonicity identity"

$$\text{(i)} \quad \rho^{2-n} \int_{B_\rho(y)} |Du|^2 - \sigma^{2-n} \int_{B_\sigma(y)} |Du|^2 = 2 \int_{B_\rho(y) \setminus B_\sigma(y)} R^{2-n} \left|\frac{\partial u}{\partial R}\right|^2,$$

valid for any $0 < \sigma < \rho < \rho_0$, provided $\overline{B}_{\rho_0}(y) \subset \Omega$, where $R = |x - y|$ and $\partial/\partial R$ means directional derivative in the radial direction $|x-y|^{-1}(x-y)$. Since it is a key tool in the study of energy minimizing maps, we give the proof of this identity.

Proof: First recall a general fact from analysis—Viz. if a_j are integrable functions on $B_{\rho_0}(y)$ and if $\int_{B_{\rho_0}(y)} \sum_{j=1}^{n} a^j D_j \zeta = 0$ for each ζ which is C^∞ with compact support in $B_{\rho_0}(y)$, then, for almost all $\rho \in (0, \rho_0)$, $\int_{B_\rho(y)} \sum_{j=1}^{n} a_j D_j \zeta = \int_{\partial B_\rho(y)} \eta \cdot a \zeta$ for any $\zeta \in C^\infty(\overline{B}_\rho(y))$, where $a = (a^1, \ldots, a^n)$ and $\eta (\equiv \rho^{-1}(x-y))$ is the outward pointing unit normal of $\partial B_\rho(y)$. (This fact is easily checked by approximating the characteristic function of the ball $B_\rho(y)$ by C^∞ functions with compact support.) Using this in the identity (v) of Section 2.2, we obtain (for almost all $\rho \in (0, \rho_0)$)

that

$$\int_{B_\rho(y)} \sum_{i,j=1}^{n} (|Du|^2 \delta_{ij} - 2D_i u \cdot D_j u) D_i \zeta^j =$$
$$= \int_{\partial B_\rho(y)} \sum_{i,j=1}^{n} (|Du|^2 \delta_{ij} - 2D_i u \cdot D_j u) \rho^{-1}(x^i - y^i) \zeta^j.$$

In this identity we choose $\zeta^j(x) := x^j - y^j$, so $D_i \zeta^j = \delta_{ij}$ and we obtain

$$(n-2) \int_{B_\rho(y)} |Du|^2 = \rho^{-1} \int_{\partial B_\rho(y)} \left(|Du|^2 - 2 \left| \frac{\partial u}{\partial R} \right|^2 \right).$$

Now by multiplying through by the factor ρ^{1-n} and noting that $\int_{\partial B_\rho} f = \frac{d}{d\rho} \int_{B_\rho} f$ for almost all ρ, we obtain the differential identity

$$\frac{d}{d\rho} \left(\rho^{2-n} \int_{B_\rho(y)} |Du|^2 \right) = 2 \frac{d}{d\rho} \left(\int_{B_\rho(y) \setminus B_\tau(y)} R^{2-n} \left| \frac{\partial u}{\partial R} \right|^2 \right)$$

for almost all $\rho \in (0, \rho_0)$ and any fixed choice of $\tau \in (0, \rho)$. Since $\int_{B_\rho} f$ is an absolutely continuous function of ρ (for any L^1-function f), we can now integrate to give the required monotonicity identity. □

Notice that since the right side of (i) is non-negative, we have in particular that

(ii) $\quad \rho^{2-n} \int_{B_\rho(y)} |Du|^2$ is an increasing function of ρ for $\rho \in (0, \rho_0)$,

and hence that the limit as $\rho \to 0$ of $\rho^{2-n} \int_{B_\rho(y)} |Du|^2$ exists; this limit is denoted $\Theta_u(y)$ and will be further discussed in the next section. An important additional conclusion, which we see by taking the limit as $\sigma \downarrow 0$ in (i), is that $\int_{B_\rho(y)} R^{2-n} \left| \frac{\partial u}{\partial R} \right|^2 < \infty$ and

(iii) $\quad \rho^{2-n} \int_{B_\rho(y)} |Du|^2 - \Theta_u(y) = 2 \int_{B_\rho(y)} R^{2-n} \left| \frac{\partial u}{\partial R} \right|^2.$

2.5 The Density Function

Definition 1 *We define the density function Θ_u of u on Ω by*

(i) $$\Theta_u(y) = \lim_{\rho \downarrow 0} \rho^{2-n} \int_{B_\rho(y)} |Du|^2.$$

As we mentioned above, this limit always exists at each point of Ω for a minimizing map u. We shall give a geometric interpretation of this below.

2.6. A Lemma of Luckhaus

For the moment, notice that the density Θ_u is upper semi-continuous on Ω; that is

(ii) $\qquad y_j \to y \in \Omega \Rightarrow \Theta_u(y) \geq \limsup_{j \to \infty} \Theta_u(y_j).$

Proof: Let $\varepsilon > 0, \rho > 0$ with $\rho + \varepsilon < \text{dist}(y, \partial\Omega)$. By the monotonicity (ii) of Section 2.4 we have $\Theta_u(y_j) \leq \rho^{2-n} \int_{B_\rho(y_j)} |Du|^2$ for j sufficiently large to ensure $\rho < \text{dist}(y_j, \partial\Omega)$. Since $B_\rho(y_j) \subset B_{\rho+\varepsilon}(y)$ for all sufficiently large j, we then have $\Theta_u(y_j) \leq \rho^{2-n} \int_{B_{\rho+\varepsilon}(y)} |Du|^2$ for all sufficiently large j, and hence we get $\limsup_{j \to \infty} \Theta_u(y_j) \leq \rho^{2-n} \int_{B_{\rho+\varepsilon}(y)} |Du|^2$. Now, by letting $\varepsilon \downarrow 0$, we conclude $\limsup_{j \to \infty} \Theta_u(y_j) \leq \rho^{2-n} \int_{B_\rho(y)} |Du|^2$, and the required inequality follows by taking the limit as $\rho \downarrow 0$. $\square$

Before we enter into the proof of the reverse Poincaré inequality for energy minimizers we need a lemma due to Luckhaus (see [Lu88] and also [Lu93]) which extends Lemma 4.3 of [SU82].

2.6 A Lemma of Luckhaus

We use the following definition:

Definition 1 (1) *If $v \in L^2(S^{n-1}; \mathbb{R}^p)$ then we say $v \in W^{1,2}(S^{n-1}; \mathbb{R}^p)$ if the homogeneous degree zero extension $\tilde{v}(r\omega) \equiv v(\omega)$, $\omega \in S^{n-1}$, $r > 0$ is in $W^{1,2}$ in some neighbourhood of S^{n-1}. (Actually if $n \geq 3$ this is the same as saying that $\tilde{v}$ is in $W^{1,2}(B_1(0); \mathbb{R}^p)$.) We say that $v \in W^{1,2}(S^{n-1}; N)$ if $v \in W^{1,2}(S^{n-1}; \mathbb{R}^p)$ and if $v(S^{n-1}) \subset N$.*

(2) *Similarly $v \in L^2(S^{n-1} \times [a,b]; \mathbb{R}^p)$ is said to be in $W^{1,2}(S^{n-1} \times [a,b]; \mathbb{R}^p)$ if the homogeneous degree zero extension of $\tilde{v}(\omega, t)$ (with respect to the S^{n-1} variable ω) is in $W^{1,2}(U \times [a,b]; \mathbb{R}^p)$ for some neighbourhood U of S^{n-1}.*

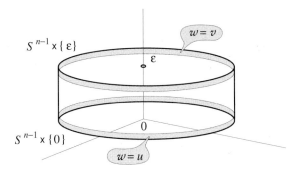

Figure 2.2: *Luckhaus' lemma*

We now state the Luckhaus lemma depicted in Figure 2.2 above.

Lemma 1 *Suppose N is an arbitrary compact subset of $\mathbb{R}^p$, $n \geq 2$, and $u, v \in W^{1,2}(S^{n-1}; N)$ in accordance with the above definition. Then for each $\varepsilon \in (0, 1)$ there is a function $w \in W^{1,2}(S^{n-1} \times [0, \varepsilon]; \mathbb{R}^p)$ such that w agrees with u in a neighbourhood of $S^{n-1} \times \{0\}$, w agrees with v in a neighbourhood of $S^{n-1} \times \{\varepsilon\}$*

$$\int_{S^{n-1} \times [0,\varepsilon]} |\overline{\nabla} w|^2 \leq C\varepsilon \int_{S^{n-1}} (|\nabla u|^2 + |\nabla v|^2) + C\varepsilon^{-1} \int_{S^{n-1}} |u - v|^2,$$

and

$$\text{dist}^2(w(x, s), N) \leq$$
$$C\varepsilon^{1-n} \left(\int_{S^{n-1}} |\nabla u|^2 + |\nabla v|^2 \right)^{1/2} \left(\int_{S^{n-1}} |u - v|^2 \right)^{1/2} + C\varepsilon^{-n} \int_{S^{n-1}} |u - v|^2$$

for a.e. $(x, s) \in S^{n-1} \times [0, \varepsilon]$. Here ∇ is the gradient on S^{n-1} and $\overline{\nabla}$ is the gradient on the product space $S^{n-1} \times [0, \varepsilon]$.

We will give the proof of this lemma in the appendix of Chapter 2.

Now we want to establish some useful corollaries of Luckhaus' Lemma.

2.7 Corollaries of Luckhaus' Lemma

First we mention the following important fact about slicing by the radial distance function:

Remark: Suppose $g \geq 0$ is integrable on $B_\rho(y)$. By virtue of the general identity $\int_{B_\rho(y) \setminus B_{\rho/2}(y)} g = \int_{\rho/2}^{\rho} (\int_{\partial B_\sigma(y)} g) \, d\sigma$, we see that for each $\theta \in (0, 1)$

(i) $$\int_{\partial B_\sigma(y)} g \leq 2\theta^{-1} \rho^{-1} \int_{B_\rho(y) \setminus B_{\rho/2}(y)} g$$

for all $\sigma \in (\frac{\rho}{2}, \rho)$ with the exception of a set of measure $\frac{\theta \rho}{2}$. (Indeed otherwise the reverse inequality would hold on a set of measure $> \frac{\theta \rho}{2}$ and by integration this would give $\int_{B_\rho(y) \setminus B_{\rho/2}(y)} g < \int_{\rho/2}^{\rho} (\int_{\partial B_\sigma(y)} g) \, d\sigma = \int_{B_\rho(y) \setminus B_{\rho/2}(y)} g$, a contradiction.)

Furthermore, if $w \in W^{1,2}(\Omega, \mathbb{R})$ (identified with some fixed chosen representative for the L^2 class of w), then for each ball $\overline{B}_\rho(y) \subset \Omega$ and each $\theta \in (0, 1)$,

(ii) $\quad w_{(\sigma)} \in W^{1,2}(S^{n-1}; \mathbb{R})$ and $\int_{S^{n-1}} |D_\omega w_{(\sigma)}|^2 d\omega \leq$

$$\leq \sigma^{3-n} \int_{\partial B_\sigma} |Dw|^2 \leq 2\theta^{-1} \rho^{2-n} \int_{B_\rho(y) \setminus B_{\rho/2}(y)} |Dw|^2$$

for all $\sigma \in (\rho/2, \rho)$, with the exception of a set of σ of measure $\frac{\theta \rho}{2}$, where $w_{(\sigma)}$ is defined by $w_{(\sigma)}(x) \equiv w(y + \sigma \omega)$, $\omega \in S^{n-1}$, and where D_ω means gradient on S^{n-1}.

2.7. Corollaries of Luckhaus' Lemma

Corollary 1 *Suppose N is a smooth compact manifold embedded in $\mathbb{R}^P$ and $\Lambda > 0$. There are $\delta_0 = \delta_0(n, N, \Lambda)$ and $C = C(n, N, \Lambda)$ such that the following hold:*

(1) If we have $\varepsilon \in (0,1)$ and if $u \in W^{1,2}(B_\rho(y); N)$ with $\rho^{2-n} \int_{B_\rho(y)} |\nabla u|^2 \leq \Lambda$, and $\varepsilon^{-2n} \rho^{-n} \int_{B_\rho(y)} |u - \lambda_{y,\rho}|^2 \leq \delta_0^2$, then there is $\sigma \in (\frac{3\rho}{4}, \rho)$ such that there is a function $w = w_\varepsilon \in W^{1,2}(B_\rho(y); N)$ which agrees with u in a neighbourhood of $\partial B_\sigma(y)$ and which satisfies

$$\sigma^{2-n} \int_{B_\sigma(y)} |Dw|^2 \leq \varepsilon \rho^{2-n} \int_{B_\rho(y)} |Du|^2 + \varepsilon^{-1} C \rho^{-n} \int_{B_\rho(y)} |u - \lambda_{y,\rho}|^2.$$

(2) If $\varepsilon \in (0, \delta_0]$, and if $u, v \in W^{1,2}(B_{(1+\varepsilon)\rho}(y) \setminus B_\rho(y); N)$ satisfy the inequalities $\rho^{2-n} \int_{B_{\rho(1+\varepsilon)}(y) \setminus B_\rho(y)} (|Du|^2 + |Dv|^2) \leq \Lambda$ and $\varepsilon^{-2n} \rho^{-n} \int_{B_{(1+\varepsilon)\rho}(y) \setminus B_\rho(y)} |u - v|^2 < \delta_0^2$, then there is $w \in W^{1,2}(B_{(1+\varepsilon)\rho}(y) \setminus B_\rho(y); N)$ such that $w = u$ in a neighbourhood of $\partial B_\rho(y)$, $w = v$ in a neighbourhood of $\partial B_{(1+\varepsilon)\rho}(y)$, and

$$\rho^{2-n} \int_{B_{(1+\varepsilon)\rho}(y) \setminus B_\rho(y)} |Dw|^2 \leq$$
$$\leq C\rho^{2-n} \int_{B_{(1+\varepsilon)\rho}(y) \setminus B_\rho(y)} (|Du|^2 + |Dv|^2) + C\varepsilon^{-2} \rho^{-n} \int_{B_{(1+\varepsilon)\rho}(y) \setminus B_\rho(y)} |u - v|^2.$$

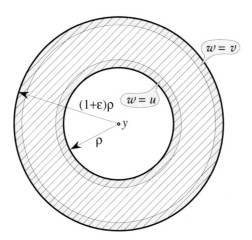

Figure 2.3: *Corollary 1(2)*

Proof (1): According to the above remark, we can choose $\sigma \in (\frac{3\rho}{4}, \rho)$ such that $u|_{\partial B_\sigma(y)} \in W^{1,2}(B_\sigma(y); N)$,

(1) $$\sigma^{3-n} \int_{\partial B_\sigma} |Du|^2 \leq C \rho^{2-n} \int_{B_\rho(y) \setminus B_{\rho/2}(y)} |Du|^2$$

and such that

$$(2) \quad \sigma^{1-n}\int_{\partial B_\sigma(y)}|u-\lambda_{y,\rho}|^2 \leq C\rho^{-n}\int_{B_\rho(y)\setminus B_{\rho/2}(y)}|u-\lambda_{y,\rho}|^2 \leq C\delta_0^2\varepsilon^{2n}.$$

Also note that, since $u(B_\rho(y)) \subset N$, $\mathrm{dist}^2(\lambda_{y,\rho}, N) \leq |u(x) - \lambda_{y,\rho}|^2$ for each $x \in B_\rho(y)$. Thus integrating over $B_\rho(y)$ we have

$$\mathrm{dist}^2(\lambda_{y,\rho}, N) \leq C\rho^{-n}\int_{B_\rho(y)}|u-\lambda_{y,\rho}|^2 \leq C\delta_0^2\varepsilon^{2n}.$$

We can thus choose $\lambda \in \mathbb{R}^p$ such that

$$(3) \quad \lambda \in N, \quad |\lambda - \lambda_{y,\rho}|^2 \leq C\rho^{-n}\int_{B_\rho(y)}|u-\lambda_{y,\rho}|^2 \leq C\delta_0^2\varepsilon^{2n}.$$

Now let $\tilde{u}$ be defined on S^{n-1} by $\tilde{u}(\omega) = u(\sigma\omega)$. Then in view of (3) we can apply Lemma 1, with $N_{C\delta_0} \equiv \{x \in \mathbb{R}^p : \mathrm{dist}(x, N) \leq C\delta_0\}$ in place of N, with $\tilde{u}$ in place of u and with $v \equiv \lambda$, in order to deduce that there is a $w_0 : S^{n-1} \to \mathbb{R}^p$ such that $w_0 = u$ in a neighbourhood of $S^{n-1} \times \{0\}$, $w_0 \equiv \lambda$ in a neighbourhood of $S^{n-1} \times \{\varepsilon\}$,

$$(4) \quad \int_{S^{n-1}\times[0,\varepsilon]}|\nabla w_0|^2 \leq C\varepsilon\int_{S^{n-1}}|\nabla\tilde{u}|^2 + C\varepsilon^{-1}\int_{S^{n-1}}|\tilde{u}-\lambda|^2$$

and also $\mathrm{dist}(w_0(x,s), N_{C\delta_0}) \leq C\delta_0^{1/2}$ by using (2), (3) in the second conclusion of Lemma 1, where $C = C(N, \Lambda, n)$. Now we can suppose (by taking a smaller $\delta_0 = \delta_0(n, N, \Lambda)$ if necessary) that $C\delta_0^{1/2} \leq \alpha$, where $\alpha > 0$ is such that the nearest point projection Π onto N is well-defined and smooth in $N_\alpha \equiv \{x \in \mathbb{R}^p : \mathrm{dist}(x, N) \leq \alpha\}$. So finally we can define $w \in W^{1,2}(B_\rho(y); N)$ by taking $w(r\omega) \equiv u(r\omega)$ for $r \in (\sigma, \rho)$, $w(r\omega) = \Pi \circ w_0(\omega, 1 - r/\sigma)$ for $r \in ((1-\varepsilon)\sigma, \sigma)$, and $w(r\omega) \equiv \lambda$ for $r \in (0, (1-\varepsilon)\sigma)$. Since $d(\Pi \circ w_0) = d\Pi_{w_0(x,s)} \circ dw_0$ it is then easy to see (with the aid of (2), (3), (4)) that this is an appropriate choice for w. So the proof of part (1) is complete.

Proof of (2): To prove part (2) we first note that by (i), (ii) above there is a set of $\sigma \in (\rho, (1 + \frac{\varepsilon}{2})\rho)$ of positive measure such that

$$(5) \quad \sigma^{3-n}\int_{\partial B_\sigma(y)}(|Du|^2 + |Dv|^2) \leq C\rho^{2-n}\varepsilon^{-1}\int_{B_{(1+\varepsilon)\rho}(y)\setminus B_\rho(y)}(|Du|^2 + |Dv|^2)$$

and

$$(6) \quad \sigma^{1-n}\int_{\partial B_\sigma(y)}|u-v|^2 \leq C\rho^{-n}\varepsilon^{-1}\int_{B_{(1+\varepsilon)\rho}(y)\setminus B_\rho(y)}|u-v|^2 \leq C\delta_0^2\varepsilon^{2n-1}.$$

Also, by (ii) we know that almost all of these σ can be selected so that $u, v \in W^{1,2}(\partial B_\sigma(y); \mathbb{R}^p)$. Now we can apply the Luckhaus lemma (with $\varepsilon/4$ in place of ε) to the functions $\tilde{u}(\omega) \equiv u(\sigma\omega)$ and $\tilde{v}(\omega) = v(\sigma\omega)$, thus giving $\tilde{w}$ on $S^{n-1} \times [0, \varepsilon/4]$

with $\widetilde{w} = \widetilde{u}$ in a neighbourhood of $S^{n-1} \times \{0\}$, $\widetilde{w} = \widetilde{v}$ in a neighbourhood of $S^{n-1} \times \{\varepsilon/4\}$,

(7)
$$\int_{S^{n-1}\times[0,\varepsilon/4]} |\overline{\nabla}\widetilde{w}|^2 \leq C\varepsilon \int_{S^{n-1}} (|\nabla\widetilde{u}|^2 + |\nabla\widetilde{v}|^2) + C\varepsilon^{-1} \int_{S^{n-1}} |\widetilde{u} - \widetilde{v}|^2$$
$$\leq C\varepsilon\sigma^{3-n} \int_{\partial B_\sigma} (|Du|^2 + |Dv|^2) + C\varepsilon^{-1}\sigma^{1-n} \int_{\partial B_\sigma} |u - v|^2$$
$$\leq C\rho^{2-n} \int_{B_{(1+\varepsilon)\rho}(y)\setminus B_\rho(y)} (|Du|^2 + |Dv|^2) + C\varepsilon^{-2}\rho^{-n} \int_{B_{(1+\varepsilon)\rho}(y)\setminus B_\rho(y)} |u - v|^2$$

by (5) and (6), and

(8)
$$\sup \operatorname{dist}^2(\widetilde{w}, N) \leq$$
$$C\left(\int_{S^{n-1}} (|\nabla\widetilde{u}|^2 + |\nabla\widetilde{v}|^2)\right)^{1/2} \left(\varepsilon^{2-2n} \int_{S^{n-1}} |\widetilde{u} - \widetilde{v}|^2\right)^{1/2} + \varepsilon^{-n}\int_{S^{n-1}} |\widetilde{u} - \widetilde{v}|^2.$$

Again by (5) and (6), the right side here is $\leq C\delta_0$ with $C = C(n, N, \Lambda)$, and so for $\delta_0 = \delta_0(n, N, \Lambda)$ small enough we conclude that $\widetilde{w}$ maps into the same neighbourhood N_α as in the proof of part (1) above. Now we can define a suitable function w, first on the ball $B_{(1+\varepsilon/2)\sigma}(y)$; we let $w(x) = \Pi \circ \widetilde{w}(\omega, r/\sigma - 1)$, with $r = |x - y| \in (\sigma, (1+\varepsilon/4)\sigma)$, $w(x) \equiv u(x)$ for $|x-y| \leq \sigma$, $w(r\omega) = v(\psi(r)\omega)$ for $r \in ((1+\varepsilon/4)\sigma, (1+\varepsilon/2)\sigma$, where $\psi(t)$ is a $C^1(\mathbb{R})$ function with the properties $\psi((1+\varepsilon/4)\sigma) = \sigma$, $\psi((1+\varepsilon/2)\sigma) = (1+\varepsilon/2)\sigma$ and $t|\psi'(t)| \leq 2$ for $t \in ((1+\varepsilon/4)\sigma, (1+\varepsilon/2)\sigma)$. In view of (7) it is straightforward to check that this satisfies the inequality stated in the lemma. □

2.8 Proof of the Reverse Poincaré Inequality

In order to complete the proof of the ε-Regularity Theorem of section 2.3 we still need to establish the reverse Poincaré inequality, i.e. the lemma

Lemma 1 *If u is energy minimizing in the sense of Section 2.1, if Λ is a given constant, and if $R^{2-n} \int_{B_R(x_0)} |Du|^2 \leq \Lambda$ for some ball $B_R(x_0)$ with closure contained in Ω, then*
$$\rho^{2-n} \int_{B_{\rho/2}(y)} |Du|^2 \leq C\rho^{-n} \int_{B_\rho(y)} |u - \lambda_{y,\rho}|^2$$
for each $y \in B_{R/2}(x_0)$, $\rho < R/4$. Here $C = C(n, N, \Lambda) > 0$.

Proof: For any $\rho \leq R/2$ and $y \in B_{R/2}(x_0)$ we have by monotonicity ((ii) of Section 2.4) that

(1) $\quad \rho^{2-n} \int_{B_\rho(y)} |Du|^2 \leq (R/2)^{2-n} \int_{B_{R/2}(y)} |Du|^2 \leq 2^{n-2} R^{2-n} \int_{B_R(x_0)} \leq 2^n \Lambda.$

Take a fixed $\rho_0 \in (0, R/2]$ and $y_0 \in B_{R/2}(x_0)$. We want to prove the inequality $\rho_0^{2-n} \int_{B_{\rho_0/2}(y_0)} |Du|^2 \leq C\rho_0^{-n} \int_{B_{\rho_0}(y_0)} |u - \lambda_{y_0,\rho_0}|^2$. By (1) there is no loss of generality in assuming

$$(2) \quad \rho_0^{-n} \int_{B_{\rho_0}(y_0)} |u - \lambda_{y_0,\rho_0}|^2 \leq \varepsilon_0^2,$$

where ε_0 is to be chosen (small) depending only on n, Λ, N (and not depending on ρ_0, R, y_0 or u); because if the reverse inequality held, then by (1) we would trivially have the required inequality with constant $C = 2^n \Lambda / \varepsilon_0^2$. So from now on we assume (2), subject to the agreement that we must eventually choose a fixed ε_0 depending only on n, N, Λ.

Let $\delta \in (0,1)$ and let $\varepsilon_0 \in (0,1)$ in (2) (depending for the moment on δ) be at least as small as $\delta^n \delta_0$, where δ_0 is as in Corollary 1 of Section 2.7. We can thus use Corollary 1(1) of Section 2.7 (with δ in place of ε) to obtain the existence of $w \in W^{1,2}(B_{\rho_0}(y_0); N)$ and $\sigma \in (\frac{3\rho_0}{4}, \rho_0)$ such that

$$\sigma^{2-n} \int_{B_\sigma(y_0)} |Dw|^2 \leq \delta \rho_0^{2-n} \int_{B_{\rho_0}(y_0)} |Du|^2 + \delta^{-1} C \rho_0^{-n} \int_{B_{\rho_0}(y_0)} |u - \lambda_{y_0,\rho_0}|^2,$$

with w agreeing with u in a neighbourhood of $\partial B_\sigma(y_0)$. Using the energy minimizing property of u on the ball $B_\sigma(y_0)$ we get

$$(3) \quad \rho_0^{2-n} \int_{B_{3\rho_0/4}(y_0)} |Du|^2 \leq \sigma^{2-n} \int_{B_\sigma(y_0)} |Dw|^2$$

$$\leq \delta \rho_0^{2-n} \int_{B_{\rho_0}(y_0)} |Du|^2 + C\delta^{-1} \rho_0^{-n} \int_{B_{\rho_0}(y_0)} |u - \lambda_{y_0,\rho_0}|^2$$

$$\leq \delta \Lambda + C\delta^{2n-1} \varepsilon_0^2.$$

Notice that then

$$(4) \quad \rho^{2-n} \int_{B_\rho(y)} |Du|^2 \leq C\delta, \quad C = C(n, N, \Lambda),$$

for any ball $B_\rho(y)$ with $y \in B_{\rho_0/2}(y_0)$ and $\rho \leq \frac{\rho_0}{4}$, because for such a ball we have $\rho^{-n} \int_{B_\rho(y)} |Du|^2 \leq (\rho_0/2)^{2-n} \int_{B_{\rho_0/4}(y)} |Du|^2 \leq 4^{n-2} \rho_0^{2-n} \int_{B_{3\rho_0/4}(y_0)} |Du|^2$ by virtue of the monotonicity formula and the inclusion $B_{\rho_0/4}(y) \subset B_{3\rho_0/4}(y_0)$. Then (keeping in mind the arbitrariness of δ) by the Poincaré inequality we deduce from (4) that

$$(5) \quad \rho^{-n} \int_{B_\rho(y)} |u - \lambda_{y,\rho}|^2, \quad \rho^{2-n} \int_{B_\rho(y)} |Du|^2 \leq \delta$$

for all such balls $B_\rho(y)$, provided only that the original inequality (2) holds with suitably small ε_0 depending only on n, N, Λ, and δ. Thus for any given $\varepsilon > 0$ and

2.8. Proof of the Reverse Poincaré Inequality

with δ chosen small enough (depending on n, N, Λ, ε), we can repeat the argument leading to (3), with ρ, y in place of ρ_0, y_0, in order to deduce

$$(6) \quad \rho^{2-n} \int_{B_{\rho/2}(y)} |Du|^2 \leq C\varepsilon \rho^{2-n} \int_{B_\rho(y)} |Du|^2 + C\varepsilon^{-1} \rho^{-n} \int_{B_\rho(y)} |u - \lambda_{y,\rho}|^2$$

provided only that $\rho \leq \rho_0/8$ and the original inequality (2) holds with ε_0 small enough, depending only on ε, N, Λ, and n. In particular, assuming such a choice of ε_0, this holds for arbitrary sub-balls $B_\sigma(z)$ with $B_{2\sigma}(z) \subset B_\rho(y)$; thus

$$\sigma^{2-n} \int_{B_{\sigma/2}(z)} |Du|^2 \leq C\varepsilon \sigma^{2-n} \int_{B_\sigma(z)} |Du|^2 + C\varepsilon^{-1} \sigma^{-n} \int_{B_\sigma(z)} |u - \lambda_{z,\sigma}|^2,$$

for each ball $B_\sigma(z)$ such that $B_{2\sigma}(z) \subset B_\rho(y)$ (and this holds for any ball $B_\rho(y)$ with $\rho \leq \rho_0/4$ and $y \in B_{\rho_0/2}(y_0)$), and hence

$$(7) \quad \sigma^2 \int_{B_{\sigma/2}(z)} |Du|^2 \leq C\varepsilon \sigma^2 \int_{B_\sigma(z)} |Du|^2 + C\varepsilon^{-1} \int_{B_\sigma(z)} |u - \lambda_{y,\rho}|^2$$

$$\leq C\varepsilon \sigma^2 \int_{B_\sigma(z)} |Du|^2 + C\varepsilon^{-1} I_1$$

where $I_1 = \int_{B_\rho(y)} |u - \lambda_{y,\rho}|^2$.

Now in view of the arbitrariness of the balls $B_\sigma(z)$ we claim that this implies the required reverse Poincaré inequality on $B_\rho(y)$. To see this we need the following abstract lemma.

Lemma 2 *Let $B_\rho(y)$ be any ball in $\mathbb{R}^n$, $k \in \mathbb{R}$, $\gamma > 0$, and let φ be any $[0, \infty)$-valued convex subadditive function on the collection of convex subsets of $B_\rho(y)$; thus $\varphi(A) \leq \sum_{j=1}^N \varphi(A_j)$ whenever $A, A_1, \ldots, A_N$ are convex subsets of $B_\rho(y)$ with $A \subset \bigcup_{j=1}^N A_j$. There is $\varepsilon_0 = \varepsilon_0(n, k)$ such that if*

$$\sigma^k \varphi(B_{\sigma/2}(z)) \leq \varepsilon_0 \sigma^k \varphi(B_\sigma(z)) + \gamma$$

whenever $B_{2\sigma}(z) \subset B_\rho(y)$, then

$$\rho^k \varphi(B_{\rho/2}(y)) \leq C\gamma, \quad C = C(n, k).$$

We give the proof of this below, but first we explain how it is used to complete the proof of the reverse Hölder inequality.

In view of (7) we may apply the lemma in the special case $\varphi(A) = \int_A |Du|^2$, $\gamma = C\varepsilon^{-1} I_1$, $C\varepsilon = \varepsilon_0$ (C as in (7)) and $k = 2$, thus giving

$$(8) \quad \rho^{2-n} \int_{B_{\rho/2}(y)} |Du|^2 \leq C\rho^{-n} \int_{B_\rho(y)} |u - \lambda_{y,\rho}|^2$$

provided $\rho \leq \rho_0/4$ and $y \in B_{\rho_0/2}(y_0)$. Finally since $B_{\rho_0/2}(y_0)$ can be covered by balls $B_{\rho_0/8}(y_j)$, $j = 1, \ldots, Q$ with $Q = Q(n)$ and $y_j \in B_{\rho_0/2}(y_0)$, it is then clear that the reverse Hölder inequality for the ball $B_{\rho_0}(y_0)$ follows by setting $\rho = \rho_0/4$ and $y = y_j$ in (8) and then summing over j. (Keeping in mind that $\int_{B_{\rho_0/4}(y_j)} |u - \lambda_{y_j, \rho_0/4}|^2 \leq \int_{B_{\rho_0}(y_0)} |u - \lambda_{y_0, \rho_0}|^2$.)

Thus to complete the proof we have only to give the proof of Lemma 2.

Proof of Lemma 2: Let

$$Q = \sup_{\{B_\sigma(z) : B_{2\sigma}(z) \subset B_\rho(y)\}} \sigma^k \varphi(B_\sigma(z)) \ (\leq (\rho/2)^k \varphi(B_\rho(y)) < \infty),$$

and then take an arbitrary ball $B_\sigma(z)$ with $B_{2\sigma}(z) \subset B_\rho(y)$. Notice that such a ball can be covered by balls $B_{\sigma/4}(z_i)$, $i = 1, \ldots, S$, with $z_i \in B_\sigma(z)$ and with $B_\sigma(z_i) \subset B_\rho(y)$; further we can evidently bound the number S by a fixed constant depending only on n.

Suppose for a moment the given inequality holds with ε in place of ε_0, with ε to be chosen depending only on n, k and not depending on γ. Then using this inequality with z_i in place of z, and $\sigma/4$ in place of σ, summing over i, and using the subadditivity of φ we have

$$\sigma^k \varphi(B_\sigma(z)) \leq 4^k \varepsilon S Q + 4^k S \gamma.$$

Taking sup on the left we thus have

$$Q \leq 4^k \varepsilon S Q + 4^k S \gamma,$$

whereupon choosing $\varepsilon = \varepsilon_0(n, k)$ such that $4^k \varepsilon_0 S \leq \frac{1}{2}$, we have

$$\sigma^k \varphi(B_\sigma(z)) \leq 4^{k+1} S \gamma,$$

for each ball $B_\sigma(z)$ with $B_{2\sigma}(z) \subset B_\rho(y)$, where C depends only on n, N, Λ. Taking $z = y$ and $\sigma = \frac{\rho}{2}$ in (7) we thus have the required conclusion with $C = 4^{k+1} S$ and $\varepsilon_0 = 1/(4^{k+1} S)$. □

2.9 The Compactness Theorem

There is also a nice compactness theorem for energy minimizing maps which is due to Luckhaus (partial results had been obtained earlier by Schoen-Uhlenbeck [SU82] and Hardt-Lin [HL87]), as follows.

Lemma 1 *If $\{u_j\}$ is a sequence of energy minimizing maps in $W^{1,2}(\Omega; N)$ with $\sup_j \int_{B_\rho(Y)} |Du_j|^2 < \infty$ for each ball $B_\rho(Y)$ with $\overline{B}_\rho(Y) \subset \Omega$, then there is a subsequence $\{u_{j'}\}$ and a minimizing harmonic map $u \in W^{1,2}(\Omega; N)$ such that $u_{j'} \to u$ in $W^{1,2}(B_\rho(y); \mathbb{R}^p)$ on each ball $\overline{B}_\rho(y) \subset \Omega$.*

2.9. The Compactness Theorem

Remarks: (1) In particular the energy $\int_{B_\rho(Y)} |Du_{j'}|^2$ converges to $\int_{B_\rho(Y)} |Du|^2$ for each ball $\overline{B}_\rho(Y) \subset \Omega$.

(2) Notice that, by the Rellich Compactness Lemma (Lemma 1 of Section 3) for bounded sequences of functions in $W^{1,2}$, there is a $W^{1,2}_{\text{loc}}(\Omega, \mathbb{R}^p)$ function u such that $u_{j'}$ converges in L^2 to u on compact subsets of Ω and $Du_{j'}$ converges locally weakly in L^2 to Du in Ω. Of course then u maps into N (in the sense that $u(x) \in N$ a.e. $x \in \Omega$) because a subsequence of the subsequence $u_{j'}$ converges pointwise a.e. to u. Thus the main content of Lemma 1 is that $Du_{j'}$ converges strongly in L^2 and that u is minimizing.

The main difficulty in proving these latter facts is that on a given ball $B_\rho(y)$ with closure contained in Ω, the values of u_j and u differ slightly near the boundary $\partial B_\rho(y)$, and so we are not able to directly use the definition of energy minimizing.

However we now have at our disposal Corollary 1(2) of Section 2.7, and this is exactly what we need to compare energies of u, u_j in a sufficiently precise manner, even though the boundary values do not coincide.

The fact that $Du_{j'}$ converges in L^2 locally on Ω is originally due to Schoen-Uhlenbeck, who used the regularity theorem to establish it. This approach however does not establish the fact that u is energy minimizing; this was not proved in full generality until the paper [Lu88].

Proof: As in the remark above, there is a subsequence $\{u_{j'}\}$ (henceforth denoted simply $\{u_j\}$) and $u \in W^{1,2}_{\text{loc}}(\Omega; N)$ such that $u_j \to u$ in L^2 and weakly in $W^{1,2}$ locally on Ω. Let $\overline{B}_{\rho_0}(y) \subset \Omega$ and let $\delta > 0$ and $\theta \in (0,1)$ be given. Choose any $M \in \{1, 2, \ldots\}$ with $\limsup \rho_0^{2-n} \int_{B_{\rho_0}(y)} |Du_j|^2 < M\delta$, and note that if $\varepsilon \in (0, (1-\theta)/M)$ we must have some integer $\ell \in \{2, \ldots, M\}$ such that

$$\rho_0^{2-n} \int_{B_{\rho_0(\theta+\ell\varepsilon)}(y) \setminus B_{\rho_0(\theta+(\ell-2)\varepsilon)}(y)} |Du_j|^2 < \delta$$

for infinitely many j, because otherwise we get that $\rho_0^{2-n} \int_{B_{\rho_0}(y)} |Du_j|^2 > M\delta$ for all sufficiently large j by summation over ℓ, contrary to the definition of M. Thus choosing such an ℓ, letting $\rho = \rho_0(\theta + (\ell-2)\varepsilon)$, and noting that $\rho(1+\varepsilon) \leq \rho_0(\theta+\ell\varepsilon) < \rho_0$, we get $\rho \in (\theta\rho_0, \rho_0)$ such that

(1) $$\rho_0^{2-n} \int_{B_{\rho(1+\varepsilon)}(y) \setminus B_\rho(y)} |Du_{j'}|^2 < \delta$$

for some subsequence j'. Of course then by weak convergence of $Du_{j'}$ to Du we also have

(2) $$\rho_0^{2-n} \int_{B_{\rho(1+\varepsilon)}(y) \setminus B_\rho(y)} |Du|^2 \leq \delta.$$

Now, by Corollary 1(2) of Section 2.7, since $\int_{B_{\rho_0}(y)} |u - u_{j'}|^2 \to 0$, for sufficiently large j' we can find $w_{j'} \in W^{1,2}(B_{\rho(1+\varepsilon)}(y) \setminus B_\rho(y); N)$ such that $w_{j'} = u$ in a neighbourhood of $\partial B_\rho(y)$, $w_{j'} = u_{j'}$ in a neighbourhood of $\partial B_{\rho(1+\varepsilon)}(y)$, and

(3) $\quad \rho^{2-n} \int_{B_{(1+\varepsilon)\rho}\setminus B_\rho(y)} |Dw_{j'}|^2 \leq$

$$\leq C\rho^{2-n} \int_{B_{(1+\varepsilon)\rho}(y)\setminus B_\rho(y)} (|Du|^2 + |Du_{j'}|^2 + \varepsilon^{-2}\rho^{-2}|u - u_{j'}|^2),$$

where C depends only on n, N. Now let $v \in W^{1,2}(B_{\theta\rho_0}(y); N)$ with $v = u$ in a neighbourhood of $\partial B_{\theta\rho}(y)$, extend v to give $\tilde{v}$ in $W^{1,2}(B_{\rho_0}(y); N)$ by taking $\tilde{v} \equiv u$ on $B_{\rho_0}(y)\setminus B_{\theta\rho_0}(y)$, and let $\tilde{u}_{j'}$ be defined by

$$\tilde{u}_j = \begin{cases} u_{j'} & \text{on } B_{\rho_0}(y) \setminus B_{(1+\varepsilon)\rho}(y) \\ w_{j'} & \text{on } B_{(1+\varepsilon)\rho}(y) \setminus B_\rho(y) \\ \tilde{v} & \text{on } B_\rho(y). \end{cases}$$

Then by the minimizing property of u_j we have

(4) $\quad \int_{B_{(1+\varepsilon)\rho}(y)} |Du_{j'}|^2 \leq \int_{B_{(1+\varepsilon)\rho}(y)} |D\tilde{u}_{j'}|^2$

$$\leq \int_{B_\rho(y)} |D\tilde{v}|^2 + \int_{B_{(1+\varepsilon)\rho}(y)\setminus B_\rho(y)} |Dw_{j'}|^2,$$

and hence by (1), (2) and (3)

(5) $\quad \rho^{2-n} \int_{B_\rho(y)} |Du|^2 \leq \liminf_{j\to\infty} \rho^{2-n} \int_{B_\rho(y)} |Du_{j'}|^2 \leq \rho^{2-n} \int_{B_\rho(y)} |D\tilde{v}|^2 + C\delta,$

where $C = C(n, N)$, and hence

$$\rho^{2-n} \int_{B_{\theta\rho_0}(y)} |Du|^2 \leq \rho^{2-n} \int_{B_{\theta\rho_0}(y)} |Dv|^2.$$

Since $\delta > 0$ was arbitrary, this shows that u is minimizing on $B_{\theta\rho_0}(y)$, and in view of the arbitrariness of θ and ρ_0, this shows that u is minimizing on all balls $B_\rho(y)$ with $\overline{B}_\rho(y) \subset \Omega$.

Finally to prove that the convergence is strong we note that if we use (5) with $v = u$, then we can conclude

$$\liminf_{j\to\infty} \rho^{2-n} \int_{B_\rho(y)} |Du_{j'}|^2 \leq \rho^{2-n} \int_{B_\rho(y)} |Du|^2 + C\delta,$$

and hence, in view of the arbitrariness of θ and δ,

$$\rho^{2-n} \liminf_{j\to\infty} \int_{B_{\rho_1}(y)} |Du_j|^2 \leq \rho^{2-n} \int_{B_{\rho_0}(y)} |Du|^2,$$

for each $\rho_1 < \rho_0$. Evidently it follows from this (keeping in mind the arbitrariness of ρ_0) that

(6) $\quad \liminf_{j\to\infty} \int_{B_\rho(y)} |Du_j|^2 \leq \int_{B_\rho(y)} |Du|^2$

for every ball $B_\rho(y)$ such that $\overline{B}_\rho(y) \subset \Omega$. Then since

$$\int_{B_\rho(y)} |Du_j - Du|^2 \equiv \int_{B_\rho(y)} |Du|^2 + \int_{B_\rho(y)} |Du_j|^2 - 2\int_{B_\rho(y)} Du \cdot Du_j,$$

we can evidently select a subsequence which converges strongly to Du on $B_\rho(y)$. Since this holds for arbitrary $\overline{B}_\rho(y) \subset \Omega$, it is then easy to see (by covering Ω by a countable collection of balls $B_{\rho_j}(y_j)$ with $\overline{B}_{\rho_j}(y_j) \subset \Omega$) that there is a subsequence such that $Du_{j'}$ converges strongly locally in all of Ω. □

2.10 Corollaries of the ε-Regularity Theorem

First we need to define the regular set reg u and the singular set sing u of u:

Definition 1 *If $u \in W^{1,2}(\Omega; \mathbb{R}^p)$, Ω an open subset of $\mathbb{R}^n$, then*

$$\operatorname{reg} u := \{x \in \Omega : u \text{ is } C^\infty \text{ in a neighbourhood of } x\}$$

is the regular set *of u, and*

$$\operatorname{sing} u := \Omega \setminus \operatorname{reg} u$$

is the singular set *of u.*

Remark: Note that by definition reg u is an open set, whereas sing u is a (relatively) closed set in Ω. We show below in Lemma 1 that sing u is small—of codimension 2. Later (in Chapter 3) we improve this even further.

Corollary 1 *There exists $\varepsilon > 0$, depending only on n, N such that if $B_\rho(y) \subset \Omega$ and if $\rho^{2-n} \int_{B_\rho(y)} |Du|^2 < \varepsilon$, then $y \in \operatorname{reg} u$ and $\sup_{B_{\rho/2}(y)} \rho^j |D^j u| \leq C$ for each $j = 0, 1, 2, \ldots$, where C depends only on j, n, N.*

Proof: The Poincaré inequality (Lemma 2 of Section 1.3) tells us that

$$\inf_{\lambda \in \mathbb{R}^p} \rho^{-n} \int_{B_\rho(y)} |u - \lambda|^2 \leq C\rho^{2-n} \int_{B_\rho(y)} |Du|^2 \leq C\varepsilon,$$

and hence Corollary 1 is a direct consequence of the ε-regularity theorem (Theorem 1 of Section 2.3). □

Now we show that the regularity theorem gives a nice way of characterizing of the regular set:

Corollary 2 $\quad \Theta_u(y) = 0 \iff y \in \operatorname{reg} u.$

Proof: "$\Leftarrow$" follows trivially from the fact that u smooth near y implies $|Du|$ is bounded near y, while "$\Rightarrow$" follows directly from Corollary 1. $\square$

The next corollary shows that the singular set of the energy-minimizing map u is actually quite small:

Lemma 1 *If $u \in W^{1,2}(\Omega; N)$ is energy minimizing in Ω then $\mathcal{H}^{n-2}(\operatorname{sing} u) = 0$. (In particular $\operatorname{sing} u = \emptyset$ in case $n = 2$.)*

Remark: Here $\mathcal{H}^{n-2}$ denotes $(n-2)$-dimensional Hausdorff measure. Thus the claim of the above lemma is that for every $\varepsilon > 0$ there exists a countable collection of balls $\{B_{\rho_j}(y_j)\}$ with $\operatorname{sing} u \subset \cup_j B_{\rho_j}(y_j)$ and $\sum_j \rho_j^{n-2} < \varepsilon$.

Proof: Let K be a compact subset of Ω, $\delta_0 < \operatorname{dist}(K, \partial\Omega)$. For $y \in \operatorname{sing} u \cap K$ we know by Corollary 1 that

$$(1) \qquad \int_{B_\rho(y)} |Du|^2 \geq \varepsilon \rho^{n-2}$$

for all $\rho < \delta_0$. For fixed $\delta < \delta_0$, pick a maximal pairwise disjoint collection of balls $B_{\delta/2}(y_j)_{j=1,\ldots,J}$ with $y_j \in K \cap \operatorname{sing} u$; that is, pick $y_j \in K \cap \operatorname{sing} u$, $j = 1, \ldots, J$, such that $B_{\delta/2}(y_j) \cap B_{\delta/2}(y_i) = \emptyset$ for all $i \neq j$ and such that J is the maximum integer such that such a collection $\{y_j\}$ exists. Then the collection $\{B_\delta(y_j)\}$ covers $K \cap \operatorname{sing} u$:

$$(2) \qquad K \cap \operatorname{sing} u \subset \cup_j B_\delta(y_j),$$

because if we could find $z \in K \cap \operatorname{sing} u \setminus (\cup_{j=1}^J B_\delta(y_j))$, then we would have pairwise disjoint balls $B_{\delta/2}(y_1), \ldots, B_{\delta/2}(y_J), B_{\delta/2}(z)$, thus contradicting the maximality of J. Using (1) with $\delta/2, y_j$ in place of ρ, y and summing over j we then have

$$(3) \qquad J\delta^{n-2} \leq 2^n \varepsilon^{-1} \int_{\cup B_\delta(y_j)} |Du|^2 \leq 2^n \varepsilon^{-1} \int_{Q_\delta} |Du|^2,$$

where $Q_\delta = \{x : \operatorname{dist}(x, K \cap \operatorname{sing} u) < \delta\}$.

In particular

$$J\delta^n \leq 2^n \delta^2 \varepsilon^{-1} \int_{Q_{\delta_0}} |Du|^2,$$

which, since $B_\delta(y_j), j = 1, \ldots, J$, cover all of $\operatorname{sing} u \cap K$, and since we can let $\delta \downarrow 0$, shows that $\operatorname{sing} u \cap K$ has Lebesgue measure zero. But then $\int_{Q_\delta} |Du|^2 \to 0$ as $\delta \downarrow 0$ by the dominated convergence theorem, and hence (3) implies that $\mathcal{H}^{n-2}(\operatorname{sing} u \cap K) = 0$. Since K was an arbitrary compact subset of Ω this shows that $\mathcal{H}^{n-2}(\operatorname{sing} u) = 0$ as required. $\square$

2.11. Remark on Upper Semicontinuity of the Density $\Theta_u(y)$

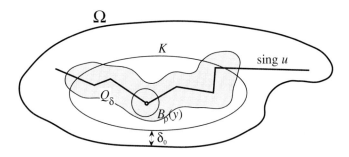

Figure 2.4: *The singular set*

2.11 A Further Remark on Upper Semicontinuity of the Density $\Theta_u(y)$

Notice that if we use the result of the compactness lemma (Lemma 1 of Section 2.9), and a very minor modification of the argument used to prove 2.5(ii), then we deduce that $\Theta_u(y)$ is actually upper-semicontinuous with respect to the joint variables u, y in the sense that if $y_j \to y \in L^2_{\text{loc}}(\Omega)$ and if $\{u_j\}$ is a sequence of energy minimizing maps from Ω into N with locally bounded energy in Ω and locally converging in L^2 (hence in $W^{1,2}$ by the compactness lemma) to u, then $\Theta_u(Y) \geq \limsup_{j \to \infty} \Theta_{u_j}(y_j)$. We use this frequently in the sequel.

2.12 Appendix to Chapter 2

In this appendix we give the proof of Luckhaus' lemma of Section 2.6. In order to prepare this proof we first mention a further property of functions with L^2 gradient:

2.12.1 Absolute Continuity Properties of Functions in $W^{1,2}$

Let Q be the cube $[a_1, b_1] \times \cdots \times [a_n, b_n]$ in $\mathbb{R}^n$ ($a_j < b_j$ real numbers, $j = 1, \ldots, n$), and let $\psi \in W^{1,2}(Q)$. Then there is a representative $\overline{\psi}$ for the L^2 class of ψ such that, for each $j = 1, \ldots, n$, $\overline{\psi}(x^1, \ldots, x^{j-1}, x^j, x^{j+1}, \ldots, x^n)$, is an absolutely continuous function of x^j for almost all fixed values of $(x^1, \ldots, x^{j-1}, x^{j+1}, \ldots, x^n)$. Here of course "almost all" is with respect to the $(n-1)$-dimensional Lebesgue measure on the $(n-1)$-dimensional cube $[a_1, b_1] \times \ldots \times [a_{j-1}, b_{j-1}] \times [a_{j+1}, b_{j+1}] \times \ldots \times [a_n, b_n]$.

Furthermore, the *classical* partial derivatives $D_j \overline{\psi}$ (defined in the usual way by $D_j \overline{\psi}(x) = \lim_{t \to 0} t^{-1}(\overline{\psi}(x + te_j) - \overline{\psi}(x))$ whenever this exists) agree a.e. with the L^2 derivatives $D_j \psi$. A discussion of these properties can be found in e.g. [GT83] or [Mo66]. We here make one further point: one procedure for constructing such a representative $\overline{\psi}$ (see e.g. [Mo66]) is to define $\overline{\psi}(x) = \lambda_x$ at all points where there exists a λ_x such that $\lim_{\rho \downarrow 0} \rho^{-n} \int_{B_\rho(y)} |\lambda_x - u(y)| \, dy = 0$, and to define $\overline{\psi}$ arbitrarily (e.g. $\overline{\psi}(x) = 0$) at points where the limit does not exist. Thus in particular if $\psi = (\psi^1, \ldots, \psi^p) : Q \to \mathbb{R}^p$, and if N is any closed subset of $\mathbb{R}^p$ with the property that $\psi(x) \in N$ for almost all $x \in \Omega$, then we can select the representative $\overline{\psi}$ to have the property $\overline{\psi}(x) \in N$ for *every* $x \in \Omega$, in addition to the absolute continuity properties mentioned above.

2.12.2 Proof of Luckhaus' Lemma (Lemma 1 of Section 2.6)

In the case $n = 2$ the functions u, v have an L^2 gradient on S^1 and hence have absolutely continuous representatives $\overline{u}$, $\overline{v}$ such that $\nabla \overline{u} = \nabla u$, $\nabla \overline{v} = \nabla v$ a.e., where $\nabla \overline{u}$, ∇u denotes respectively the classical gradient and the weak gradient of $\overline{u}$ on S^1. Furthermore, by 1-dimensional calculus on S^1 and by the Cauchy-Schwarz inequality we have

$$\sup_{S^1} |\overline{u} - \overline{v}|^2 \leq \int_{S^1} |\nabla |\overline{u} - \overline{v}|^2| + (2\pi)^{-1} \int_{S^1} |\overline{u} - \overline{v}|^2$$
$$\leq C \big(\int_{S^1} |\nabla(\overline{u} - \overline{v})|^2 \big)^{1/2} \big(\int_{S^1} |\overline{u} - \overline{v}|^2 \big)^{1/2} + C \int_{S^1} |\overline{u} - \overline{v}|^2,$$

If we now define

$$w(\omega, s) = \overline{u}(\omega) + \frac{s}{\varepsilon} \big(\overline{v}(\omega) - \overline{u}(\omega) \big),$$

then, letting $\overline{\nabla} w$ denote the gradient of w on $S^1 \times [0, \varepsilon]$, we have

$$|\overline{\nabla} w| \leq |\nabla \overline{u}| + |\nabla(\overline{v} - \overline{u})| + \frac{1}{\varepsilon} |\overline{v} - \overline{u}|,$$

and hence

$$|\overline{\nabla} w|^2 \leq 8(|\nabla \overline{u}|^2 + |\nabla \overline{v}|^2) + 2\varepsilon^{-2} |\overline{v} - \overline{u}|^2.$$

By integrating this over $S^1 \times [0, \varepsilon]$, we get the first claim of the Luckhaus Lemma. Further, since $\overline{u}(S^1) \subset N$, the above inequality for $\sup_{S^1} |\overline{u} - \overline{v}|$ implies that for each $\omega \in S^1$, $s \in [0, \varepsilon]$, we have

$$\text{dist}(w(\omega, s), N) \leq C \big(\int_{S^1} |\nabla(\overline{u} - \overline{v})|^2 \big)^{1/4} \big(\int_{S^1} |\overline{u} - \overline{v}|^2 \big)^{1/4} + C \big(\int_{S^1} |\overline{u} - \overline{v}|^2 \big)^{1/2},$$

which is the second claim of the Luckhaus Lemma (indeed it is stronger since in this case $n = 2$ we get no ε dependence on the right). This completes the proof in the case $n = 2$, so from now on assume $n \geq 3$.

Again choose representatives $\overline{u}$, $\overline{v}$ for u, v which have homogeneous degree zero extensions to $\mathbb{R}^n$ having the absolute continuity properties of $\overline{\psi}$ of 2.12.1 on the cube $[0, 1] \times \cdots \times [0, 1]$.

Without changing notation, we also let $\overline{u}$, $\overline{v}$ denote these homogeneous degree zero extensions on $\mathbb{R}^n$; thus $\overline{u}(r\omega) \equiv \overline{u}(\omega)$ for $r > 0$ and $\omega \in S^{n-1}$, and $D\overline{u} = \nabla u$ on

2.12. Appendix to Chapter 2

S^{n-1}, $D\bar{v} = \nabla v$ on S^{n-1}. Notice also that $D\bar{u}(x) = |x|^{-1} D\bar{u}(\omega)$, with $\omega = |x|^{-1}x$, and hence (since $n \geq 3$) we have in particular that

(1) $$\int_{[-1,1]^n} (|D\bar{u}|^2 + |D\bar{v}|^2) \leq C \int_{S^{n-1}} (|\nabla u|^2 + |\nabla v|^2),$$

and also of course

(2) $$\int_{[-1,1]^n} |\bar{u} - \bar{v}|^2 \leq C \int_{S^{n-1}} |u - v|^2.$$

Now for $\varepsilon \in (0, \frac{1}{8})$ and $i = (i_1, \ldots, i_n) \in \mathbb{Z}^n$ ($\mathbb{Z} = \{0, \pm 1, \pm 2, \ldots\}$), we let $Q_{i,\varepsilon}$ denote the cube $[i_1 \varepsilon, (i_1 + 1)\varepsilon] \times \cdots \times [i_n \varepsilon, (i_n + 1)\varepsilon]$, and for a given non-negative measurable function $f : [-1, 1]^n \to \mathbb{R}$ we let $\tilde{f} : Q_{i,\varepsilon} \to \mathbb{R}$ be defined by

$$\tilde{f}(x) = \sum_{\{i : Q_{i,\varepsilon} \subset [-\frac{1}{2}, \frac{1}{2}]^n\}} f(x + \varepsilon i), \quad x \in \tilde{Q}_{0,\varepsilon}, \quad \tilde{Q}_{0,\varepsilon} = \cup_{|j| \leq 1} Q_{j,\varepsilon}.$$

Then

$$\int_{Q_{0,\varepsilon}} \tilde{f}(x) \, dx = \int_{\cup \{Q_{i,\varepsilon} : Q_{i,\varepsilon} \subset [-\frac{1}{2}, \frac{1}{2}]^n\}} f(x) \, dx \leq \int_{[-1,1]^n} f(x) \, dx,$$

and hence for any $K \geq 1$ we have

$$\varepsilon^n \sum_{\{i : Q_{i,\varepsilon} \subset [-\frac{1}{2}, \frac{1}{2}]^n\}} f(x + \varepsilon i) \leq K \int_{[-1,1]^n} f(x) \, dx$$

for all $x \in Q_{0,\varepsilon}$ with the exception of a set of measure $\leq C\varepsilon^n/K$, where $C = C(n)$. Similarly, since by Fubini's theorem

$$\int_{Q_{0,\varepsilon}} \int_0^\varepsilon \tilde{f}(x + te_n) \, dt \, dx \leq \varepsilon \int_{\tilde{Q}_{0,\varepsilon}} \tilde{f}(x) \, dx \leq \varepsilon \int_{[-1,1]^n} f(x) \, dx, \quad \tilde{Q}_{0,\varepsilon} = \cup_{|j| \leq 1} Q_{j,\varepsilon},$$

we have

$$\varepsilon^{n-1} \sum_{\{i : Q_{i,\varepsilon} \subset [-\frac{1}{2}, \frac{1}{2}]^n\}} \int_0^\varepsilon f(x + te_n + \varepsilon i) \, dt \leq K \int_{[-1,1]^n} f(x) \, dx$$

for all $x \in Q_{0,\varepsilon}$ with the exception of a set of measure $\leq C\varepsilon^n/K$, and generally, for any $\ell \in \{0, \ldots, n\}$,

$$\varepsilon^{n-\ell} \sum_{\{i : Q_{i,\varepsilon} \subset [-\frac{1}{2}, \frac{1}{2}]^n\}} \int_{F^{(\ell)}} f(x + y + \varepsilon i) \, d\mathcal{H}^\ell(y) \leq K \int_{[-1,1]^n} f(x) \, dx$$

for all ℓ-faces $F^{(\ell)}$ of $Q_{0,\varepsilon}$ and all $x \in Q_{0,\varepsilon}$ with the exception of a set of measure $\leq C\varepsilon^n/K$. Notice that this last inequality implies

(3) $$\varepsilon^{n-\ell} \sum_{\{i : Q_{i,\varepsilon} \subset [-\frac{1}{2}, \frac{1}{2}]^n\}} \sum_{\ell\text{-faces } F^{(\ell)} \text{ of } x + Q_{i,\varepsilon}} \int_{F^{(\ell)}} f(y) \, d\mathcal{H}^\ell(y) \leq K \int_{[-1,1]^n} f(x) \, dx$$

for all $\ell \in \{0, \ldots, n\}$ and all $x \in Q_{0,\varepsilon}$ with the exception of a set of measure $\leq C\varepsilon^n/K$.

Now by the absolute continuity properties of Appendix 2.12.1, we can select representatives $\bar{u}, \bar{v}, \overline{Du}, \overline{Dv}$, for the L^2 classes of u, v, Du, Dv such that for almost all $x \in Q_{0,\varepsilon}$ all of the functions $\bar{u}, \bar{v}, \overline{Du}, \overline{Du}$ are defined $\mathcal{H}^\ell$-a.e. on each of the ℓ-dimensional faces F^ℓ of each of the cubes $x + Q_{i,\varepsilon}$ with $Q_{i,\varepsilon} \subset [-\frac{1}{2}, \frac{1}{2}]^n$ for each $\ell = 0, \ldots, n$ and, furthermore, such that on each such ℓ-dimensional face, $\bar{u}, \bar{v}$ have L^2-gradients which coincide $\mathcal{H}^\ell$-a.e. with the tangential parts of $\overline{Du}, \overline{Dv}$. Now by applying (3) with $f = |\bar{u} - \bar{v}|^2$ and $f = |\overline{Du}|^2 + |\overline{Dv}|^2$ we see that we can select $x = a \in Q_{0,\varepsilon}$ such that the above properties hold and also such that

$$(4) \quad \varepsilon^{n-\ell} \sum_{\{i : Q_{i,\varepsilon} \subset [-\frac{1}{2}, \frac{1}{2}]^n\}} \sum_{\ell\text{-faces } F^{(\ell)} \text{ of } a+Q_{i,\varepsilon}} \int_{F^{(\ell)}} f(y) \, d\mathcal{H}^\ell(y) \leq$$

$$\leq C \int_{[-1,1]^n} f(x) \, dx \text{ with } f = |\bar{u} - \bar{v}|^2 \text{ or } f = |\overline{Du}|^2 + |\overline{Dv}|^2,$$

for each $\ell \in \{0, \ldots, n\}$, where $C = C(n)$.

Next, let Q be any one of the cubes $a + Q_{i,\varepsilon}$ with $Q_{i,\varepsilon} \subset [-\frac{1}{2}, \frac{1}{2}]^n$, and we proceed to define a $W^{1,2}$ function $w = w^{(i,\varepsilon)}$ on $Q \times [0, \varepsilon]$ which agrees with $\bar{u}$ on $Q \times \{0\}$ at all points of Q where $\bar{u}$ exists, agrees with $\bar{v}$ on $Q \times \{\varepsilon\}$ at all points of Q where $\bar{v}$ exists, and which is such that

$$(5) \quad \int_{Q \times [0,\varepsilon]} |\overline{D}w|^2 \leq C \sum_{j=1}^{n-1} \varepsilon^{n-j+1} \sum_{\text{all } j\text{-faces } F^{(j)} \text{ of } Q} \int_{F^{(j)}} (|\overline{Du}|^2 + |\overline{Dv}|^2)$$

$$+ C\varepsilon^{n-2} \sum_{\text{all 1-faces } F^{(1)}} \int_{F^{(1)}} |u - v|^2$$

with $\overline{D} =$ gradient on $Q \times [0, \varepsilon]$, and

$$(6) \quad \text{dist}^2(w(x), N) \leq$$

$$C \max_{\text{1-faces } F^{(1)} \text{ of } Q} \left(\left(\int_{F^{(1)}} |D(\bar{u} - \bar{v})|^2 \right)^{1/2} \left(\int_{F^{(1)}} |\bar{u} - \bar{v}|^2 \right)^{1/2} + C\varepsilon^{-1} \int_{F^{(1)}} |\bar{u} - \bar{v}|^2 \right).$$

Let E be any one of the edges (i.e. 1-dimensional faces) of Q. By 1-dimensional calculus along the line segment E, we have (since the length of E is ε)

$$(7) \quad \sup_E |\bar{u} - \bar{v}|^2 \leq \int_E |D|\bar{u} - \bar{v}|^2| + \varepsilon^{-1} \int_E |\bar{u} - \bar{v}|^2.$$

Hence by using the Cauchy-Schwarz inequality we obtain

$$(8) \quad \sup_E |\bar{u} - \bar{v}|^2 \leq 2 \left(\int_E |D(\bar{u} - \bar{v})|^2 \right)^{1/2} \left(\int_E |\bar{u} - \bar{v}|^2 \right)^{1/2} + \varepsilon^{-1} \int_E |\bar{u} - \bar{v}|^2$$

2.12. Appendix to Chapter 2

Then we can define an $\mathbb{R}^p$-valued function w on $Q \times [0, \varepsilon]$ by the following inductive procedure. We first define w on $Q \times \{0\}$ and $Q \times \{\varepsilon\}$ by

(9) $$w(x, 0) = \overline{u}(x), \quad w(x, \varepsilon) = \overline{v}(x), \quad x \in Q.$$

Next we extend w to each $F^{(1)} \times [0, \varepsilon]$, where $F^{(1)}$ is any 1-dimensional face (i.e. edge) of Q, by defining

$$w(x, s) = (1 - \frac{s}{\varepsilon})\overline{u}(x) + \frac{s}{\varepsilon}\overline{v}(x), \quad x \in F^{(1)}, \ s \in [0, \varepsilon].$$

Notice that then by (8) and the fact that $\overline{u}(\mathbb{R}^n) \subset N$ we have

$$\operatorname{dist}^2(w(x, s), N) \leq \max_{\text{1-faces } F^{(1)} \text{ of } Q} \sup_{F^{(1)}} |\overline{v} - \overline{u}|^2 \leq$$

$$2 \max_{\text{1-faces } F^{(1)} \text{ of } Q} \left(\left(\int_{F^{(1)}} |D(\overline{u} - \overline{v})|^2 \right)^{1/2} \left(\int_{F^{(1)}} |\overline{u} - \overline{v}|^2 \right)^{1/2} + \varepsilon^{-1} \int_{F^{(1)}} |\overline{u} - \overline{v}|^2 \right).$$

Also notice that by direct computation

(11) $$\sup_{s \in [0,\varepsilon]} |\overline{D}w(x, s)|^2 \leq 8(|D\overline{u}(x)|^2 + |D\overline{v}(x)|^2) + \frac{2}{\varepsilon^2}|\overline{u}(x) - \overline{v}(x)|^2$$

at any x in any edge $F^{(1)}$ of Q (where $\overline{D} = (\frac{\partial}{\partial x}, \frac{\partial}{\partial s})$, $\frac{\partial}{\partial x} = $ gradient on $F^{(1)}$), hence

(12) $$\int_{F^{(1)} \times [0,\varepsilon]} |\overline{D}w|^2 \leq C\varepsilon \int_{F^{(1)}} (|D\overline{u}|^2 + |D\overline{v}|^2) + C\varepsilon^{-1} \int_{F^{(1)}} |\overline{u} - \overline{v}|^2.$$

For $\ell \geq 2$ we now proceed inductively by homogeneous extension into faces of larger and larger dimension. More precisely, assume $\ell \geq 2$, and that w is already defined (with L^2 gradient) on all $F^{(\ell-1)} \times [0, \varepsilon]$ and $w(x, 0) \equiv \overline{u}(x)$, $w(x, \varepsilon) \equiv \overline{v}(x)$ on $F^{(\ell)}$. Since $\partial(F^{(\ell)} \times [0, \varepsilon])$ is the union of $F^{(\ell-1)} \times [0, \varepsilon]$ (over the $\ell - 1$ faces $F^{(\ell-1)}$ of $F^{(\ell)}$) together with $F^{(\ell)} \times \{0\}$ and $F^{(\ell)} \times \{\varepsilon\}$, we then have that w is already well defined $\mathcal{H}^\ell$-a.e. on $\partial(F^{(\ell)} \times [0, \varepsilon])$. We can thus use homogeneous degree zero extension of $w|\partial(F^{(\ell)} \times [0, \varepsilon])$ into $F^{(\ell)} \times [0, \varepsilon]$ with origin at the point $(q, \varepsilon/2)$, where q is the center point of $F^{(\ell)}$. Then by direct computation we have

(13) $$\int_{F^{(\ell)} \times [0,\varepsilon]} |\overline{D}w|^2 \leq C\varepsilon \int_{F^{(\ell)}} (|D\overline{u}|^2 + |D\overline{v}|^2) + C\varepsilon \sum_{\text{all } F^{(\ell-1)}} \int_{F^{(\ell-1)} \times [0,\varepsilon]} |\overline{D}w|^2,$$

where $\overline{D} = (\frac{\partial}{\partial x}, \frac{\partial}{\partial s})$, $\frac{\partial}{\partial x} = $ gradient on $F^{(\ell)}$ on the left and on $F^{(\ell-1)}$ on the right. (In checking the ε dependence here it suffices to check the inequality only in the special case $\varepsilon = 1$, because we deduce the general case from this case by the scaling $(x, s) \mapsto (\varepsilon x, \varepsilon s)$.) So by mathematical induction based on (13) we conclude that, for all $\ell \in \{2, \ldots, n\}$, w can be extended to all of $F^{(\ell)} \times [0, \varepsilon]$ ($F^{(\ell)} = $ any ℓ-face of Q) such that w has L^2 gradient $\overline{D}w$ on all $F^{(\ell)} \times [0, \varepsilon]$ with

$$\int_{F^{(\ell)} \times [0,\varepsilon]} |\overline{D}w|^2 \leq C\varepsilon^{\ell-1} \sum_{\text{all 1-faces } F^{(1)} \text{ of } Q} \int_{F^{(1)} \times [0,\varepsilon]} |\overline{D}w|^2$$

$$+ C \sum_{j=1}^{\ell} \varepsilon^{\ell-j+1} \sum_{\text{all } j\text{-faces } F^{(j)} \text{ of } Q} \int_{F^{(j)}} (|D\overline{u}|^2 + |D\overline{v}|^2).$$

Furthermore notice that homogeneous degree zero extension preserves the bound (10). Thus by (10), (12), (14) (with $\ell = n$) we conclude the existence of $w : Q \to \mathbb{R}^p$ as in (5) and (6). Since $Q = a + Q_{i,\varepsilon}$, we should write $w = w^{(i,\varepsilon)}$. Since the construction of $w^{(i,\varepsilon)}$ is such as to ensure that $w^{(i,\varepsilon)} = w^{(j,\varepsilon)}$ $\mathcal{H}^{\ell+1}$-a.e. on $F^{(\ell)} \times [0, \varepsilon]$ for any common ℓ-face $F^{(\ell)}$ of two different cubes $a + Q_{i,\varepsilon}$ and $a + Q_{j,\varepsilon}$, we can then define a $W^{1,2}$ function w on $[-\frac{1}{4}, \frac{1}{4}]^n$ by setting $w(x,s) = w^{(i,\varepsilon)}(x,s)$ for $(x,s) \in (a + Q_{i,\varepsilon}) \times [0, \varepsilon]$. (We are assuming $\varepsilon \in (0, \frac{1}{8})$ and $a \in Q_{0,\varepsilon}$, hence we automatically have that $[-\frac{1}{4}, \frac{1}{4}]^n \subset \cup \{Q_{i,\varepsilon} : a + Q_{i,\varepsilon} \subset [-\frac{1}{2}, \frac{1}{2}]^n\}$.) Notice that then by summing over i in (5) (keeping in mind that $Q = a + Q_{i,\varepsilon}$) and using (4) and (6) we get

$$(15) \quad \int_{[-\frac{1}{4},\frac{1}{4}]^n \times [0,\varepsilon]} |\overline{D}w|^2 \leq C\varepsilon \int_{[-1,1]^n} (|Du|^2 + |Dv|^2) + C\varepsilon^{-1} \int_{[-1,1]^n} |u - v|^2$$

and

$$\operatorname{dist}^2(w(x), N) \leq$$

$$\leq C \max_{\text{1-faces } F^{(1)} \in \mathcal{F}} \left(\int_{F^{(1)}} |D(\overline{u} - \overline{v})|^2 \int_{F^{(1)}} |\overline{u} - \overline{v}|^2 \right)^{1/2} + C\varepsilon^{-1} \int_{F^{(1)}} |\overline{u} - \overline{v}|^2$$

$$\leq C\varepsilon^{1-n} \left(\int_{[-1,1]^n} (|Du|^2 + |Dv|^2) \int_{[-1,1]^n} |u - v|^2 \right)^{1/2} + C\varepsilon^{-n} \int_{[-1,1]^n} |u - v|^2,$$

where $\mathcal{F}$ denotes the collection of all 1-faces of cubes in the collection $\{a + Q_{i,\varepsilon} : Q_{i,\varepsilon} \subset [-\frac{1}{2}, \frac{1}{2}]^n\}$.

Defining $\widetilde{w} = w|([-\frac{1}{4}, \frac{1}{4}]^n \setminus [-\frac{1}{8}, \frac{1}{8}]^n) \times [0, \varepsilon]$, and, using 2.12.1 and Fubini's Theorem, we can choose $\rho \in [\frac{1}{8}, \frac{1}{4}]$ such that w has L^2-gradient on $\partial([-\rho, \rho]^n) \times [0, \varepsilon]$ and such that

$$\int_{\partial([-\rho,\rho]^n) \times [0,\varepsilon]} |\overline{D}w|^2 \leq C \int_{[-\frac{1}{4},\frac{1}{4}]^n \times [0,\varepsilon]} |\overline{D}w|^2.$$

Then finally let Ψ be the radial map from 0 taking S^{n-1} to $\partial([-\rho, \rho]^n)$ (notice that this is a Lipschitz piecewise C^1 map with a Lipschitz piecewise C^1 inverse). Thus we can define $\hat{w}$ on $S^{n-1} \times [0, \varepsilon]$ by $\hat{w}(\omega, s) = \widetilde{w}(\Psi(\omega), s)$, and one then readily checks that this map $\hat{w}$ has the properties claimed for w in the statement of the Luckhaus lemma. (In particular, since $w(x, 0) \equiv \overline{u}(x)$ for $x \in \partial([-\rho, \rho]^n)$ and since $\overline{u}$ is homogeneous of degree zero in $\mathbb{R}^n$, we then have by definition that $\hat{w}(\omega, 0) \equiv \overline{u}(\omega)$ a.e. on S^{n-1}. Similarly $\hat{w}(\omega, \varepsilon) \equiv \overline{v}(\omega)$ a.e. on S^{n-1}.)

This completes the proof of the Luckhaus lemma.

2.12.3 Nearest point projection

Here we want to give a proof of the fact that if N is a compact C^∞ (resp. C^ω) manifold which is isometrically embedded in $\mathbb{R}^p$, then there is a tubular neighbourhood $U = \{x \in \mathbb{R}^p : \operatorname{dist}(x, N) < \delta\}$ of N such that the nearest point projection map

2.12. Appendix to Chapter 2

(taking a point $x \in U$ to the nearest point of N) is well defined and C^∞ (resp. C^ω). At the same time we want to discuss the geometrical significance of the induced linear map and the Hessian of this nearest point projection.

The main results are described in the following theorem:

Theorem 1 *If N is a compact C^∞ (resp. C^ω) submanifold of dimension q embedded in $\mathbb{R}^p$, then there is $\delta = \delta(N) > 0$ and a map $\Pi \in C^\infty(\{x : \text{dist}(x, N) < \delta\}; \mathbb{R}^p)$ (resp. $\Pi \in C^\omega(\{x : \text{dist}(x, N) < \delta\}; \mathbb{R}^p)$) such that the following properties hold:*

(i) $\begin{cases} \Pi(y) \in N, \quad y - \Pi(y) \in T^\perp_{\Pi(y)} N \quad |\Pi(y) - y| = \text{dist}(y, N), \text{ and} \\ |z - y| > \text{dist}(y, N) \text{ for any } z \in N \setminus \{\Pi(y)\} \end{cases}$

for all $y \in \mathbb{R}^p$ with $\text{dist}(y, N) < \delta$,

(ii) $\quad \Pi(y + z) \equiv y, \quad \text{for } y \in N, \, z \in (T_y N)^\perp, \, |z| < \delta,$

(iii) $\quad D_v \Pi|_y \equiv p_{\Pi(y)}(v), \quad v \in \mathbb{R}^p, \, \text{dist}(y, N) < \delta,$

where D_v denotes directional derivative $v \cdot D$ and $p_{\Pi(y)}$ denotes orthogonal projection of $\mathbb{R}^p$ onto $T_{\Pi(y)} N$,

(iv) $\quad v_1 \cdot \text{Hess}\, \Pi_y(v_2, v_3) = \frac{1}{2} \sum v^\perp_{\sigma_1} \cdot \text{Hess}\, \Pi_y(v^\top_{\sigma_2}, v^\top_{\sigma_3}),$
for $\text{dist}(y, N) < \delta$, $v_1, v_2, v_3 \in \mathbb{R}^p$,

where $v^\top|_y = p_{\Pi(y)} v$, $v^\perp = v - v^\top$; the sum on the right is over all 6 permutations $\sigma_1, \sigma_2, \sigma_3$ of $1, 2, 3$ and $\text{Hess}\, \Pi_y$ denotes the Hessian of Π at y (thus $\text{Hess}\, \Pi_y$ is a symmetric map $\mathbb{R}^p \times \mathbb{R}^p \to \mathbb{R}^p$),

(v) $\quad \text{Hess}\, \Pi_y(v_1, v_2) = -A_y(v_1, v_2), \quad y \in N, \quad v_1, v_2 \in T_y N,$

where A_y is the second fundamental form of N at y. Furthermore, if $u : \Omega \to N$ is a smooth map ($\Omega \subset \mathbb{R}^n$ open), then

(vi) $\quad (\text{Hess}\, u|_x(v_1, v_2))^{\perp u(x)} \equiv \text{Hess}\, \Pi_{u(x)}(D_{v_1} u(x), D_{v_2} u(x))$
$\qquad\qquad\qquad\qquad\quad \equiv -A_{u(x)}(D_{v_1} u(x), D_{v_2} u(x)),$

for $x \in \Omega$, $v_1, v_2 \in \mathbb{R}^n$, where $v^{\perp u(x)}$ means $v^\perp|_{u(x)}$.

Remark: Notice that in particular (iv) implies that $v_1 \cdot \text{Hess}\, \Pi_y(v_2, v_3)$ is a symmetric function of $(v_1, v_2, v_3) \in \mathbb{R}^p \times \mathbb{R}^p \times \mathbb{R}^p$ and that $\text{Hess}\, \Pi_y(v^\perp_1, v^\perp_2) \equiv 0$ for all $v_1, v_2 \in \mathbb{R}^p$; indeed $v_1 \cdot \text{Hess}\, \Pi_y(v_2, v_3) = 0$ whenever at least 2 of v_1, v_2, v_3 are in $(T_{\Pi(y)} N)^\perp$.

Proof: We describe the proof in the C^ω case. The proof for the C^∞ case is identical, using smooth maps rather than real-analytic at each stage. (If N is merely C^k for some $k \geq 2$, then the proof here shows that the nearest point projection is C^{k-1}.)

Thus assume N is isometrically embedded in $\mathbb{R}^p$ and is a real-analytic manifold. This means that for each $y_0 \in N$ we can find real-analytic functions $u : W \to (T_{y_0}N)^\perp$, where $q < p$ is the dimension of N and W is a neighbourhood of y_0 in the affine space $y_0 + T_{y_0}N$, such that

$$\operatorname{graph} u := \{x + u(x) : x \in W\}$$

is a neighbourhood V of y_0 in N.

For notational convenience we can assume

$$y_0 = 0, \quad T_{y_0}N = \mathbb{R}^q \times \{0\}, \quad (T_{y_0}N)^\perp = \{0\} \times \mathbb{R}^{p-q}.$$

W should then be reinterpreted as a neighbourhood of 0 in $\mathbb{R}^q$ rather than $\mathbb{R}^q \times \{0\}$, and $u = (u^1, \ldots, u^{p-q}) : W \to \mathbb{R}^{p-q}$ with $D_j u(0) = 0$, $j = 1, \ldots, q$. Define a map $\varphi : W \times \mathbb{R}^{p-q} \to \mathbb{R}^p$ by

$$\varphi(x) = (x, u(x)),$$

so that φ is a real analytic diffeomorphism of W onto the neighbourhood $V \subset N$. Notice also that, at each point $x \in W$, $D_1\varphi(x), \ldots, D_q\varphi(x)$ are a basis for the tangent space $T_{\varphi(x)}N$. Hence by the Gram-Schmidt orthogonalization process we can construct real-analytic functions $\nu_1, \ldots, \nu_{p-q}$ on W such that $\nu_1(x), \ldots, \nu_{p-q}(x)$ are an orthonormal basis for $(T_{\varphi(x)}N)^\perp$ at each point $x \in W$ and $\nu_1(0), \ldots, \nu_{p-q}(0)$ are the standard basis vectors $e_{q+1}, \ldots, e_p$. Then we define a map $\Phi : W \times \mathbb{R}^{p-q} \to \mathbb{R}^p$ by

$$\Phi(x, y) = \varphi(x) + \sum_{j=1}^{p-q} y^j \nu_j(x).$$

By direct computation we then have

(1) $$d\Phi|_{(0,0)} = \mathbf{1}_{\mathbb{R}^p}.$$

Also, by construction Φ is real analytic on $W \times \mathbb{R}^{p-q}$, hence using local power series expansions we can extend Φ to give a holomorphic mapping $\widetilde{\Phi}$ of the complex variables $z^1, \ldots, z^p$ in some neighbourhood of $W \times \mathbb{C}^{p-q}$ (thinking now of W as a subset of $\mathbb{R}^q \times \{0\} \subset \mathbb{C}^q$ which Φ maps into $\mathbb{R}^p \subset \mathbb{C}^p$). Now of course the identity (1) guarantees that $d\widetilde{\Phi}|_0$ is the identity (as a complex linear map) of $\mathbb{C}^p$ onto $\mathbb{C}^p$, and hence the holomorphic inverse function theorem implies that there are complex neighbourhoods W_1, W_2 of 0 in $\mathbb{C}^p$ such that $\widetilde{\Phi}$ is a holomorphic map of W_1 onto W_2 with holomorphic inverse. But this evidently implies in particular that $\Phi|W_1 \cap (\mathbb{R}^p \times \{0\})$ is a (real) analytic map onto an open subset $\widetilde{V}$ of $\mathbb{R}^p$, with $0 \in \widetilde{V}$, having a real-analytic inverse. Thus in particular, for suitable $\delta = \delta(N) > 0$, Φ gives a real-analytic diffeomorphism of a neighbourhood of $\overline{B}^q_\delta(0) \times \overline{B}^{p-q}_\delta(0)$ onto some neighbourhood of 0 such that $|D\Phi|, |D\Phi^{-1}| \leq C$ on $\overline{B}^q_\delta(0) \times \overline{B}^{p-q}_\delta(0)$ and $\Phi(\overline{B}^q_\delta(0) \times \overline{B}^{p-q}_\delta(0))$ respectively, where $C = C(N)$. Now for $\theta \in (0,1)$ to be chosen shortly, take any $z \in \Phi(B^q_{\delta/2}(0) \times B^{p-q}_{\theta\delta}(0))$; say $z = \Phi(\xi, \eta)$, $(\xi, \eta) \in B^q_{\delta/2} \times B^{p-q}_{\theta\delta}$. Evidently, since $\varphi(\xi) = \Phi(\xi, 0)$, we then have $|\varphi(\xi) - z| \leq C\eta \leq C\theta\delta$, $C = C(N)$, so in particular $\operatorname{dist}(z, N) \leq C\theta\delta$, while on the other hand $\operatorname{dist}(z, \partial\Phi(B_\delta \times B^{p-q}_\delta)) \geq$

2.12. Appendix to Chapter 2

$C^{-1}\delta$ for suitable $C = C(N)$. Therefore, if $\theta = \theta(N) \in (0,1)$ is chosen small enough, we have

$$\min_{x \in \overline{B}_\delta^q} |\varphi(x) - z| < \delta, \tag{2}$$

and the minimum is attained only at interior points $x \in B_\delta^q(0)$. At any such point x we then have

$$\langle D_j\varphi(x), \varphi(x) - z \rangle = 0, \quad j = 1, \ldots, q,$$

and since $D_j\varphi(x)$, $j = 1, \ldots, q$, are a basis for $T_{\varphi(x)}N$, it therefore follows that $z - \varphi(x) \in (T_{\varphi(x)}N)^\perp$ at any point $x \in \overline{B}_\delta^q(0)$ where the minimum in (2) is attained; thus

$$z = \varphi(x) + \sum_j \lambda^j \nu_j(x) \text{ for suitable } \lambda = (\lambda^1, \ldots, \lambda^{p-q}) \text{ with } |\lambda| < \delta. \tag{3}$$

But now this says precisely $\Phi(x, \lambda) = \Phi(\xi, \eta)$ with both (x, λ) and (ξ, η) in $B_\delta^q(0) \times B_\delta^{p-q}(0)$, which contradicts that fact that Φ is one-to-one on $B_\delta^q(0) \times B_\delta^{p-q}(0)$ unless $x = \xi$ and $\lambda = \eta$.

Thus (changing notation to the extent that we write δ in place of $\theta\delta$) we have proved that, for $\delta = \delta(N) \in (0, \frac{1}{2})$ small enough, each point $z = \Phi(x, y) \in \Phi(B_\delta^q(0) \times B_\delta^{p-q}(0))$ has the point $\varphi(x)$ as unique nearest point projection $\Pi(z)$ onto N, $z - \varphi(x) \in (T_{\varphi(x)}N)^\perp$, and, since Φ is a real-analytic diffeomorphism of $B_\delta^q(0) \times B_\delta^{p-q}(0)$, this nearest point projection $\Pi : z = \Phi(x, y) \mapsto \varphi(x)$ is a real-analytic map. Indeed Π is given explicitly in $\Phi(B_\delta^q(0) \times B_\delta^{p-q}(0))$ by

$$\Pi = \varphi \circ P \circ \Phi^{-1}, \tag{4}$$

where P is the orthogonal projection of $\mathbb{R}^p$ onto $\mathbb{R}^q \times \{0\}$. Since N is compact and y_0 was an arbitrary point of N to begin with, this completes the proof of the existence of a real analytic Π satisfying (i), (ii) for suitable $\delta = \delta(N) > 0$.

Next we introduce the notation

$$v^\top|_y = D_v\Pi|_y, \quad v^\perp = v - v^\top, \quad v \in \mathbb{R}^p, \; y \in U, \tag{5}$$

where D_v denotes the directional derivative $v \cdot D$ in $\mathbb{R}^p$ and where $U = \{x \in \mathbb{R}^p : \text{dist}(x, N) < \delta\}$. Since Π is the identity on N and since $\Pi(y + t\eta) \equiv y$ for $y \in N$ and $\eta \in (T_y N)^\perp$ with $|t\eta| < \delta$, we see that geometrically $v^\top$ is just the tangential part of v relative to N at the nearest point $\Pi(y)$; that is, (iii) holds.

By applying the directional derivative operator D_{v_1} to the first identity in (5) (with $v = v_2 \in C^\infty(U; \mathbb{R}^p)$), we deduce that

$$\text{Hess } \Pi_y(v_1(y), v_2(y)) = (D_{v_1}v_2^\top - (D_{v_1}v_2)^\top)|_y, \quad y \in U, \; v_1, v_2 \in C^\infty(U; \mathbb{R}^p). \tag{6}$$

Taking $v_1, v_3^\perp$ and then $v_1^\top, v_2^\top$ in place of the pair v_1, v_2 we in particular deduce the two facts that

$$\begin{aligned} \text{Hess } \Pi_y(v_1^\top, v_3^\perp) &= -(D_{v_1^\top}v_3^\perp)^\top \in T_{\Pi(y)}N \\ \text{Hess } \Pi_y(v_1^\top, v_2^\top) &= (D_{v_1^\top}v_2^\top)^\perp \in (T_{\Pi(y)}N)^\perp. \end{aligned} \tag{7}$$

Since $v_2^\top \cdot (D_{v_1^\top} v_3^\perp) \equiv -v_3^\perp \cdot D_{v_1^\top} v_2^\top$ (which one checks by applying $D_{v_1^\top}$ to the identity $v_2^\top \cdot v_3^\perp \equiv 0$), this gives in turn that

(8) $\qquad v_2^\top \cdot \text{Hess}\, \Pi_y(v_1^\top, v_3^\perp) = v_3^\perp \cdot \text{Hess}\, \Pi_y(v_1^\top, v_2^\top)$ for $y \in U$.

Also, from (i) and (ii) we deduce that $\Pi(y + s\eta_1 + t\eta_2) \equiv \Pi\big(\Pi(y) + (y - \Pi(y)) + s\eta_1 + t\eta_2\big) \equiv \Pi(y)$ for $\eta_1, \eta_2 \in (T_{\Pi(y)}N)^\perp$ and for $|s|, |t|$ small enough, and hence $\frac{\partial^2 \Pi(y+s\eta_1+t\eta_2)}{\partial s \partial t}\big|_{s=t=0} = 0$; that is,

(9) $\qquad \text{Hess}\, \Pi_y(\eta_1, \eta_2) = 0, \quad y \in U, \ \eta_1, \eta_2 \in (T_{\Pi(y)}N)^\perp.$

Since $v \equiv v^\top + v^\perp$ on U (for any given $v \in \mathbb{R}^p$), using linearity together with (6), (7), (8), and (9) it is now straightforward to check the identity (iv).

Next recall that the second fundamental form A_y of N at $y \in N$ is the symmetric bilinear form on $T_y N$ with values in $(T_y N)^\perp$ defined by

$$A_y(v_1^\top, v_2^\top) = (D_{v_1^\top} v_2^\top)^\perp|_y, \quad v_1, v_2 \in \mathbb{R}^p, \ y \in N.$$

Using (6) with $v_1^\top, v_2^\top$ in place of v_1, v_2 we get

$$A_y(v_1^\top, v_2^\top) \equiv -\text{Hess}\, \Pi_y(v_1^\top, v_2^\top)$$

for $v_1, v_2 \in \mathbb{R}^p, y \in N$, and hence (v) is proved.

Finally, if $u : \Omega \to N$ is smooth ($\Omega \subset \mathbb{R}^n$ open), then $D_j u(x) \in T_{u(x)} N$ for $x \in \Omega$ and $j = 1, \ldots, n$, and since $u(x) \equiv \Pi(u(x))$, and hence $D_k D_j u(x) = \text{Hess}_{u(x)}(D_k u(x), D_j u(x)) = d\Pi_{u(x)}(D_k D_j u(x))$, we deduce

$$(D_k D_j u(x))^\perp \equiv \text{Hess}\, \Pi_{u(x)}(D_j u(x), D_k u(x)) \equiv -A_{u(x)}(D_k u(x), D_j u(x)), \quad x \in \Omega,$$

by (v), so (vi) is proved. $\qquad\square$

2.12.4 Proof of the ε-regularity theorem in case $n = 2$

Here we assume $n = 2$, that $u \in W^{1,2}(\Omega; N)$ is energy minimizing, and that $\overline{B_R(y)} \subset \Omega$.

Let $\Lambda > 0$ be any constant such that

$$\int_{B_R(y)} |Du|^2 \leq \Lambda.$$

As we mentioned in Section 2.3, in the present case $n = 2$ we obtain without further hypotheses that

$$R^j \sup_{B_{R/2}(y)} |D^j u| \leq C, \quad C = C(j, N, \Lambda), \ j = 0, 1, \ldots.$$

2.12. Appendix to Chapter 2

The hard step in the proof (as it is for $n \geq 3$ also) involves showing that $u \in C^{1,\alpha}$. We do this here—the remainder of the proof proceeds, using Schauder theory for linear equations, exactly as described in Section 2.3.

We first prove that $u \in C^{0,\alpha}(B_{R/2}(0))$ for some $\alpha = \alpha(\Lambda, N) \in (0, 1)$.

Note that by the absolute continuity properties of functions with gradient in L^2 (see the discussion in Appendix 2.12 above), there is a representative $\bar{u}$ of the L^2 class of u such that $\bar{u}|\partial B_\tau(y)$ is an absolutely continuous function for almost all $\tau \in (0, R]$, $\bar{u}(\partial B_\tau(y)) \subset N$, $D\bar{u}$ (the classical gradient of the function $\bar{u}$) exists a.e. and agrees a.e. with the L^2 gradient Du of u.

Now take $\sigma \in (0, R]$. By the formula $\int_{B_\sigma(y) \setminus B_{\sigma/2}(y)} |Du|^2 = \int_{\sigma/2}^{\sigma} \int_{\partial B_\tau(y)} |Du|^2 \, d\tau$, we know that, for any $\theta \in (0, \frac{1}{2})$,

$$\int_{\partial B_\tau(y)} |D\bar{u}|^2 \leq \theta^{-1} \sigma^{-1} \int_{B_\sigma(y) \setminus B_{\sigma/2}(y)} |D\bar{u}|^2$$

for all $\tau \in (\frac{\sigma}{2}, \sigma)$ with the exception of a set of 1-dimensional Lebesgue measure $\leq \theta \sigma$. Taking $\theta = \frac{1}{4}$ and such a τ with $\bar{u}|\partial B_\tau(y)$ absolutely continuous, we then have

$$(1) \quad \int_{\partial B_\tau(y)} |D\bar{u}|^2 \leq 4\sigma^{-1} \int_{B_\sigma(y) \setminus B_{\sigma/2}(y)} |D\bar{u}|^2.$$

On the other hand, since $\bar{u}|\partial B_\tau(y)$ is absolutely continuous, we can use 1-dimensional calculus on the circle $\partial B_\tau(y)$ and the Cauchy-Schwarz inequality to give

$$(2) \quad \sup_{x_1, x_2 \in \partial B_\tau(y)} |\bar{u}(x_1) - \bar{u}(x_2)| \leq \int_{\partial B_\tau(y)} |D\bar{u}| \leq \sqrt{2\pi\tau} \left(\int_{\partial B_\tau(y)} |D\bar{u}|^2 \right)^{\frac{1}{2}}.$$

Hence by (1) we have, except for a set of $\tau \in (\frac{\sigma}{2}, \sigma)$ of measure $\leq \frac{\sigma}{4}$,

$$(3) \quad \sup_{x, x_0 \in \partial B_\tau(y)} |\bar{u}(x) - \bar{u}(x_0)| \leq 6 \left(\int_{B_\sigma(y) \setminus B_{\sigma/2}(y)} |Du|^2 \right)^{\frac{1}{2}}.$$

Now let $\delta = \delta(N) > 0$ be small enough to ensure that the nearest point map Π (taking a point in $\mathbb{R}^P$ to the nearest point of N) is well-defined and smooth on the set $\{x \in \mathbb{R}^P : \text{dist}(x, N) \leq \delta\}$ (see Appendix 2.12.3 above), and let $\varepsilon \in (0, \frac{1}{2})$ be for the moment arbitrary. If $6 \left(\int_{B_\sigma(y) \setminus B_{\sigma/2}(y)} |Du|^2 \right)^{1/2} \leq \varepsilon$ and if $x_0 \in \partial B_\tau(y)$, then, provided ε is sufficiently small depending only on N and Λ, we have from (3) that the homogeneous degree 1 extension $\tilde{u} : B_\tau(y) \to \mathbb{R}^P$ defined by

$$(4) \quad \tilde{u}(y + r\omega) = \bar{u}(x_0) + \tau^{-1} r(\bar{u}(y + \tau\omega) - \bar{u}(x_0)), \quad \omega \in S^1, \, r \in (0, \tau],$$

remains in the δ-neighbourhood of N, and hence we can define

$$\hat{u} = \Pi \circ \tilde{u}$$

on $\overline{B}_\tau(y)$. Then $\widehat{u}$ agrees with $\overline{u}$ on $\partial B_\tau(y)$ and by (4), (2), and (1)

$$\mathcal{E}_{B_\tau(y)}(\widehat{u}) \leq C\tau \int_{\partial B_\tau(y)} |D\overline{u}|^2 \leq C \int_{B_\sigma(y)\setminus B_{\sigma/2}(y)} |Du|^2.$$

But by definition of energy minimizing we have $\mathcal{E}_{B_\tau(y)}(u) \leq \mathcal{E}_{B_\tau(y)}(\widehat{u})$, and hence this gives

(5) $$\int_{B_{\sigma/2}(y)} |Du|^2 \leq C \int_{B_\sigma(y)\setminus B_{\sigma/2}(y)} |Du|^2.$$

Keep in mind that this has been proved so far only under the assumption that

(6) $$\int_{B_\sigma(y)\setminus B_{\sigma/2}(y)} |Du|^2 \leq \varepsilon,$$

with ε sufficiently small (depending on N). On the other hand for each Q we have $B_R(y) = B_{R/2^Q}(y) \cup (\cup_{j=1}^Q (B_{R/2^{j-1}}(y) \setminus B_{R/2^j}(y)))$ and hence, if we take Q depending on ε, there is at least one j such that (6) holds with $R/2^j$ in place of σ and then (5) gives therefore that

$$\int_{B_{\gamma R}(y)} |Du|^2 \leq C\varepsilon,$$

where $\gamma \in (0,1)$ depends only on Λ, ε and N. With ε/C in place of ε and selecting $\varepsilon \leq \varepsilon_0$, $\varepsilon_0 = \varepsilon_0(N,\Lambda)$ small enough, we thus deduce from (5) that

(7) $$\int_{B_{\rho/2}(y)} |Du|^2 \leq C \int_{B_\rho(y)\setminus B_{\rho/2}(y)} |Du|^2 \leq \varepsilon$$

for any $\rho \leq \gamma R$, provided that $\mathcal{E}_{B_R(y)}(u) \leq \Lambda$, where $\gamma = \gamma(N,\Lambda,\varepsilon) \in (0,\frac{1}{2})$. By adding $C\int_{B_{\rho/2}(y)} |Du|^2$ to each side of this inequality we get

$$\int_{B_{\rho/2}(y)} |Du|^2 \leq \theta \int_{B_\rho(y)} |Du|^2, \quad \rho \leq \gamma R,$$

where $\theta = \frac{C}{1+C} \in (0,1)$ depends only on n, N, Λ, ε. By iteration this gives that $\int_{B_{\gamma R/2^j}} |Du|^2 \leq C 2^{-\alpha j}$ for each $j = 1, 2, \ldots$, and hence (since any $\rho \in (0, \gamma R]$ lies in some interval $(\gamma R/2^j, \gamma R/2^{j-1}]$ for some $j \geq 1$) we have

$$\int_{B_\sigma(y)} |Du|^2 \leq C\left(\frac{\sigma}{R}\right)^\alpha, \quad \sigma \in (0,R), \quad C = C(\Lambda, N).$$

(Notice that we can arrange for the inequality to hold trivially for $\sigma \in (\gamma R, R)$ by choosing C suitably large.) On the other hand $\mathcal{E}_{B_R(y)}(u) \leq \Lambda$ implies $\mathcal{E}_{B_{R/2}(z)}(u) \leq \Lambda$ for any $z \in B_{R/2}(y)$, so the above actually implies that

$$\int_{B_\sigma(z)} |Du|^2 \leq C\left(\frac{\sigma}{R}\right)^\alpha, \quad \sigma \in (0,R], \quad z \in B_{R/2}(y).$$

2.12. Appendix to Chapter 2

But now Morrey's lemma (Lemma 3 of Section 1.3) implies that $u \in C^{0,\alpha}(B_{R/2}(y))$ and that
$$[u]_{\alpha, B_{R/2}(y)} \leq C,$$
with C depending only on Λ, N and α; here $\alpha = \alpha(N, \Lambda) \in (0, 1)$.

Next we show that this holds for *every* $\alpha \in (0, 1)$. To see this we let $\rho \in (0, \gamma R]$ be arbitrary, and note that, by (7) and the first part of the argument above, for any $\varepsilon \in (0, \frac{1}{2})$ and for suitable $\gamma = \gamma(\varepsilon, N, \Lambda) \in (0, \frac{1}{2})$, there is $\tau \in (\frac{\rho}{2}, \rho)$ such that (3) holds, and hence

$$\tag{8} \sup_{\partial B_\tau(y)} |\bar{u} - \bar{u}(x_0)| \leq C\varepsilon^{1/2},$$

where x_0 is any point of $\partial B_\tau(y)$. As in the last part of the proof of Lemma 1 of Section 1.8, we can find a harmonic function $v \in W^{1,2}(B_\tau(y); \mathbb{R}^p) \cap C^0(\overline{B}_\tau(y); \mathbb{R}^p)$ with $v = \bar{u}$ on ∂B_τ and with

$$\tag{9} \max_{\overline{B}_\tau(y)} |v - \bar{u}(x_0)| \leq C\varepsilon^{1/2}.$$

Then

$$\tag{10} \int_{B_\sigma(y)} |Du|^2 \leq 2 \int_{B_\sigma(y)} |D(u-v)|^2 + 2 \int_{B_\sigma(y)} |Dv|^2$$
$$\leq 2 \int_{B_\tau(y)} |D(u-v)|^2 + 2 \frac{\sigma^2}{\tau^2} \int_{B_\tau(y)} |Dv|^2$$

for any $\sigma \in (0, \tau]$, where we used the fact that $|Dv|^2$ is a subharmonic function and hence $\sigma^{-2} \int_{B_\sigma(y)} |Dv|^2$ is an increasing function of $\sigma \in (0, \tau]$. (See e.g. [GT83].)

Now on the other hand by the inequality (11) in the proof of Lemma 1 of Section 1.8, and by (8), (9), we know that

$$\int_{B_\tau(y)} |D(u-v)|^2 \leq C \max_{\partial B_\tau(y)} |u - v| \int_{B_\tau(y)} |Du|^2$$
$$\leq C \left(\max_{\partial B_\tau(y)} |\bar{u} - \bar{u}(x_0)| + \max_{\partial B_\tau(y)} |v - \bar{u}(x_0)| \right) \int_{B_\tau(y)} |Du|^2$$
$$\leq C\varepsilon^{1/2} \int_{B_\rho(y)} |Du|^2.$$

So (10) implies

$$\int_{B_\sigma(y)} |Du|^2 \leq C\varepsilon^{1/2} \int_{B_\rho(y)} |Du|^2 + 2\frac{\sigma^2}{\tau^2} \int_{B_\tau(y)} |Dv|^2.$$

Also since v is harmonic and agrees with $\bar{u}$ on $\partial B_\tau(y)$, we have $\int_{B_\tau(y)} |Dv|^2 \leq \int_{B_\tau(y)} |Du|^2$, and hence this gives

$$\int_{B_\sigma(y)} |Du|^2 \leq \left(C\varepsilon^{1/2} + 2\frac{\sigma^2}{\tau^2} \right) \int_{B_\rho(y)} |Du|^2, \quad \sigma \in (0, \tau].$$

Since $\tau \in (\frac{\rho}{2}, \rho)$ we thus have

$$(11) \quad \int_{B_\sigma(y)} |Du|^2 \leq \left(C\varepsilon^{1/2} + 4\frac{\sigma^2}{\rho^2} \right) \int_{B_\rho(y)} |Du|^2, \quad \sigma \in (0, \tfrac{\rho}{2}]$$

for any $\varepsilon \in (0, \frac{1}{2})$, provided $\rho \leq \gamma R$ with $\gamma = \gamma(N, \Lambda, \varepsilon) \in (0, \frac{1}{2})$ is sufficiently small. Notice that for any given $\alpha \in (0, 1)$ we can select a $\theta = \theta(\alpha) \in (0, 1)$ such that $4\theta^2 \leq \theta^{2\alpha}/2$ and then choose $\varepsilon = \varepsilon(\alpha, N, \Lambda)$ such that $C\varepsilon \leq \theta^{2\alpha}/2$, with C as in (11). So, if we take $\sigma = \theta\rho$ with these choices, (11) gives

$$\int_{B_{\theta\rho}(y)} |Du|^2 \leq \theta^{2\alpha} \int_{B_\rho(y)} |Du|^2, \quad \rho \in (0, \gamma R],$$

where $\gamma = \gamma(N, \Lambda, \alpha)$. Of course the same argument applies in any balls $B_\rho(z)$ with $\rho \leq \gamma R$ and $z \in B_{\gamma R}(y)$, with $\gamma = \gamma(\Lambda, N) \in (0, \frac{1}{2}]$ sufficiently small, so we have actually shown

$$(12) \quad \int_{B_{\theta\rho}(z)} |Du|^2 \leq \theta^{2\alpha} \int_{B_\rho(z)} |Du|^2, \quad \sigma \in (0, \gamma R], \ z \in B_{\gamma R}(y),$$

with $\gamma = \gamma(N, \Lambda, \alpha)$. Iterating (12), we easily check that it implies

$$\int_{B_\sigma(z)} |Du|^2 \leq C \left(\frac{\sigma}{\rho} \right)^{2\alpha} \int_{B_\rho(z)} |Du|^2, \quad \sigma \in (0, \rho], \ \rho \in (0, \gamma R], \ z \in B_{\gamma R}(y).$$

By the Morrey lemma (Lemma 3 of Section 1.3) we then have $u \in C^{0,\alpha}(B_{\gamma R}(y))$ and

$$|u(x_1) - u(x_2)| \leq C \left(\frac{|x_1 - x_2|}{R} \right)^\alpha, \quad x_1, x_2 \in B_{\gamma R}(y),$$

for suitable $\gamma = \gamma(\alpha, N, \Lambda) \in (0, \frac{1}{2})$. On the other hand the same argument applies starting with any ball $B_{R/2}(z)$ ($z \in B_{R/2}(y)$) in place of $B_R(y)$, so that we have actually proved

$$|u(x_1) - u(x_2)| \leq C \left(\frac{|x_1 - x_2|}{R} \right)^\alpha, \quad x_1, x_2 \in B_{R/2}(y),$$

with $C = C(\Lambda, N, \alpha)$, as required.

The proof that u is $C^{1,\alpha}$ now follows exactly the last part of the argument in the proof of Lemma 1 of Section 1.8. $\square$

Chapter 3

Approximation Properties of the Singular Set

In this chapter u continues to denote an energy minimizing map of Ω into N, with Ω an open subset of $\mathbb{R}^n$.

3.1 Definition of Tangent Map

Let $B_{\rho_0}(y)$ with $\overline{B}_{\rho_0}(y) \subset \Omega$, and for any $\rho > 0$ consider the scaled function $u_{y,\rho}$ defined by

$$u_{y,\rho}(x) = u(y + \rho x).$$

Notice that $u_{y,\rho}$ is well-defined on the ball $B_{\rho_0}(0)$; furthermore, if $\sigma > 0$ is arbitrary and $\rho < \frac{\rho_0}{\sigma}$, we have (using $Du_{y,\rho}(x) = \rho(Du)(y + \rho x)$, and making a change of variable $\tilde{x} = y + \rho x$ in the energy integral for $u_{y,\rho}$)

(i) $\quad \sigma^{2-n} \int_{B_\sigma(0)} |Du_{y,\rho}|^2 = (\sigma\rho)^{2-n} \int_{B_{\sigma\rho}(y)} |Du|^2 \leq \rho_0^{2-n} \int_{B_{\rho_0}(y)} |Du|^2,$

where in the last inequality we used monotonicity ((ii) of Section 2.4). Thus if $\rho_j \downarrow 0$ then $\limsup_{j \to \infty} \int_{B_\sigma(0)} |Du_{y,\rho_j}|^2 < \infty$ for each $\sigma > 0$, and hence by the compactness theorem (Lemma 1 of Section 2.9) there is a subsequence $\rho_{j'}$ such that $u_{y,\rho_{j'}} \to \varphi$ locally in $\mathbb{R}^n$ with respect to the $W^{1,2}$-norm, where $\varphi : \mathbb{R}^n \to N$ is an energy minimizing map (in the sense of Section 2.1) with $\Omega = \mathbb{R}^n$. Any φ which is obtained in this way is called a *tangent map of u at y*; further properties of tangent maps are discussed below. In general it is *not* true that such tangent maps need be unique (see [Wh92])—that is, if we choose different sequences ρ_j (or different subsequences $\rho_{j'}$) then we may get a different limit map. In case the target N is real analytic rather than merely C^∞, it remains an open question whether or not we do or do not have uniqueness of φ. In Section 3.10 it will be shown that if N is real analytic and if one of the tangent maps φ satisfies $\operatorname{sing}\varphi = \{0\}$, then φ is the unique tangent map.

3.2 Properties of Tangent Maps

Let $\rho_j \downarrow 0$ be one of the sequences such that the re-scaled maps $u_{y,\rho_j} \to \varphi$ as described above. Since u_{y,ρ_j} converges in energy to φ, we have, after setting $\rho = \rho_j$ and taking limits on each side of 3.1(i) as $j \to \infty$,

$$\sigma^{2-n} \int_{B_\sigma(0)} |D\varphi|^2 = \Theta_u(y),$$

where we used the definition $\Theta_u(y) = \lim_{\rho \downarrow 0} \rho^{2-n} \int_{B_\rho(y)} |Du|^2$. Thus in particular $\sigma^{2-n} \int_{B_\sigma(0)} |D\varphi|^2$ is a constant function of σ, and, since by definition $\Theta_\varphi(0) = \lim_{\sigma \downarrow 0} \sigma^{2-n} \int_{B_\sigma(0)} |D\varphi|^2$, we have

(i) $$\Theta_u(y) = \Theta_\varphi(0) \equiv \sigma^{2-n} \int_{B_\sigma(0)} |D\varphi|^2 \quad \forall \sigma > 0.$$

Thus any tangent map of u at y has scaled energy constant and equal to the density of u at y; this is also a nice interpretation of the density of u at y.

Furthermore if we apply the monotonicity formula (iii) of Section 2.4 to φ then we get the identity

$$0 = \sigma^{2-n} \int_{B_\sigma(0)} |D\varphi|^2 - \tau^{2-n} \int_{B_\tau(0)} |D\varphi|^2 = \int_{B_\sigma(0) \setminus B_\tau(0)} R^{2-n} \left|\frac{\partial \varphi}{\partial R}\right|^2,$$

so that $\partial\varphi/\partial R = 0$ a.e., and since $\varphi \in W^{1,2}_{\text{loc}}(\mathbb{R}^n; \mathbb{R}^p)$ it is correct to conclude from this, by integration along rays, that

(ii) $$\varphi(\lambda x) \equiv \varphi(x) \quad \forall \lambda > 0,\, x \in \mathbb{R}^n.$$

This is a key property of tangent maps, and enables us to use the further properties of homogeneous degree zero minimizers (see Section 3.3 below) in studying them.

We conclude this section with another nice characterization of the regular set of u:

(iii) $\qquad y \in \operatorname{reg} u \iff \exists$ a *constant* tangent map φ of u at y.

To prove (iii), note that by Corollary 2 of Section 2.10 we have $y \in \operatorname{reg} u \iff \Theta_u(y) = 0$, but $\Theta_u(y) = 0 \iff \varphi \equiv \text{const.}$ by (i).

3.3 Properties of Homogeneous Degree Zero Minimizers

Suppose $\varphi : \mathbb{R}^n \to N$ is a homogeneous degree zero minimizer (e.g. a tangent map of u at some point y); thus $\varphi(\lambda x) \equiv \varphi(x)$ for all $\lambda > 0$, $x \in \mathbb{R}^n$.

3.3. Properties of Homogeneous Degree Zero Minimizers

We first observe that the density $\Theta_\varphi(y)$ is maximum at $y = 0$; in fact, by the monotonicity formula of Section 2.4, for each $\rho > 0$ and each $y \in \mathbb{R}^n$

$$2 \int_{B_\rho(y)} R_y^{2-n} \left| \frac{\partial \varphi}{\partial R_y} \right|^2 + \Theta_\varphi(y) = \rho^{2-n} \int_{B_\rho(y)} |D\varphi|^2,$$

where $R_y(x) \equiv |x - y|$ and $\partial/\partial R_y = |x-y|^{-1}(x-y) \cdot D$. Now $B_\rho(y) \subset B_{\rho+|y|}(0)$, so that

$$\rho^{2-n} \int_{B_\rho(y)} |D\varphi|^2 \le \rho^{2-n} \int_{B_{\rho+|y|}(0)} |D\varphi|^2$$

$$= \left(1 + \frac{|y|}{\rho}\right)^{n-2} (\rho + |y|)^{2-n} \int_{B_{\rho+|y|}(0)} |D\varphi|^2$$

$$\equiv \left(1 + \frac{|y|}{\rho}\right)^{n-2} \Theta_\varphi(0),$$

because φ is homogeneous of degree zero (which guarantees that $\tau^{2-n} \int_{B_\tau(0)} |D\varphi|^2 \equiv \Theta_\varphi(0)$). Thus letting $\rho \uparrow \infty$, we get

$$2 \int_{\mathbb{R}^n} R_y^{2-n} \left| \frac{\partial \varphi}{\partial R_y} \right|^2 + \Theta_\varphi(y) \le \Theta_\varphi(0),$$

which establishes the required inequality

(i) $$\Theta_\varphi(y) \le \Theta_\varphi(0).$$

Notice also that this argument shows that *equality* in (i) implies $\partial \varphi/\partial R_y = 0$ a.e.; that is, $\varphi(y + \lambda x) \equiv \varphi(y + x)$ for each $\lambda > 0$. Since we also have (by assumption) $\varphi(x) \equiv \varphi(\lambda x)$ we can then compute for any $\lambda > 0$ and $x \in \mathbb{R}^n$ that

$$\varphi(x) = \varphi(\lambda x) = \varphi(y + (\lambda x - y)) = \varphi(y + \lambda^{-2}(\lambda x - y))$$
$$= \varphi(\lambda(y + \lambda^{-2}(\lambda x - y))) = \varphi(x + ty),$$

where $t = \lambda - \lambda^{-1}$ is an arbitrary real number. So let $S(\varphi)$ be defined by

(ii) $$S(\varphi) = \{y \in \mathbb{R}^n : \Theta_\varphi(y) = \Theta_\varphi(0)\}.$$

Then we have shown that $\varphi(x) \equiv \varphi(x + ty)$ for all $x \in \mathbb{R}^n$, $t \in \mathbb{R}$, and $y \in S(\varphi)$. Then of course $\varphi(x + az_1 + bz_2) \equiv \varphi(x)$ for all $a, b \in \mathbb{R}$ and $z_1, z_2 \in S(\varphi)$. But if $z \in \mathbb{R}^n$ and $\varphi(x + z) \equiv \varphi(x)$ for all $x \in \mathbb{R}^n$, then trivially $\Theta_\varphi(z) = \Theta_\varphi(0)$ (and hence $z \in S(\varphi)$ by definition of $S(\varphi)$), so we conclude

$S(\varphi)$ is a linear subspace of $\mathbb{R}^n$ and $\varphi(x + y) \equiv \varphi(x)$, $x \in \mathbb{R}^n$, $y \in S(\varphi)$.

(Thus φ is invariant under composition with translation by elements of $S(\varphi)$.) Notice of course that

(iii) $$\dim S(\varphi) = n \iff S(\varphi) = \mathbb{R}^n \iff \varphi = \text{const.}$$

Also, a homogeneous degree zero map which is not constant clearly cannot be continuous at 0, so we always have $0 \in \text{sing}\, \varphi$ if φ is non-constant, and hence, since $\varphi(x + z) \equiv \varphi(x)$ for any $z \in S(\varphi)$, we have

(iv) $$S(\varphi) \subset \text{sing}\, \varphi$$

for any non-constant homogeneous degree zero minimizer φ.

3.4 Further Properties of sing u

For any $y \in \Omega$ and any tangent map φ of u at y we shall let $S(\varphi)$ be the linear subspace of points y such that $\Theta_\varphi(y) = \Theta_\varphi(0)$, as discussed in the previous section. Notice that then by 3.2(iii) we have

(i) $y \in \operatorname{sing} u \iff \dim S(\varphi) \leq n-1$ for every tangent map φ of u at y.

Now for each $j = 0, 1, \ldots, n-1$ we define

$$\mathcal{S}_j = \{y \in \operatorname{sing} u : \dim S(\varphi) \leq j \text{ for } \textit{all} \text{ tangent maps } \varphi \text{ of } u \text{ at } y\}.$$

Then we have

(ii) $\mathcal{S}_0 \subset \mathcal{S}_1 \subset \cdots \subset \mathcal{S}_{n-3} = \mathcal{S}_{n-2} = \mathcal{S}_{n-1} = \operatorname{sing} u.$

To see this first note that $\mathcal{S}_{n-1} = \operatorname{sing} u$ is just (i), and the inclusion $\mathcal{S}_{j-1} \subset \mathcal{S}_j$ is true by definition. Also, if $\mathcal{S}_{n-3}$ is not equal to both $\mathcal{S}_{n-2}$ and $\mathcal{S}_{n-1}$, then we can find $y \in \operatorname{sing} u$ at which there is a tangent map φ with $\dim S(\varphi) = n-1$ or $n-2$; but then $\mathcal{H}^{n-2}(S(\varphi)) = \infty$ and hence (since $S(\varphi) \subset \operatorname{sing} \varphi$ by 3.3(iv)) we have $\mathcal{H}^{n-2}(\operatorname{sing} \varphi) = \infty$, contradicting the fact that $\mathcal{H}^{n-2}(\operatorname{sing} \varphi) = 0$ by Lemma 1 of Section 2.10.

The subsets $\mathcal{S}_j$ are mainly important because of the following lemma, which is a direct modification of the corresponding result for minimal surfaces by F. Almgren [Ag83]; the lemma can be thought of as a refinement of the "dimension reducing" argument of Federer [FH69] (for this see also the discussion in the appendix of [Si83a]):

Lemma 1 *For each $j = 0, \ldots, n-3$, $\dim \mathcal{S}_j \leq j$, and, for each $\alpha > 0$, $\mathcal{S}_0 \cap \{x : \Theta_u(x) = \alpha\}$ is a discrete set.*

Remark: Here "dim" means Hausdorff dimension; thus $\dim \mathcal{S}_j \leq j$ means simply that $\mathcal{H}^{j+\varepsilon}(\mathcal{S}_j) = 0$ for each $\varepsilon > 0$.

Before we give the proof of this lemma, we note the following corollary.

Corollary 1 $\dim \operatorname{sing} u \leq n-3$, *and if N is a 2-dimensional surface of genus ≥ 1, then $\dim \operatorname{sing} u \leq n-4$. More generally, if all tangent maps $\varphi \in W^{1,2}_{\text{loc}}(\mathbb{R}^m; N)$ of u satisfy $\dim S(\varphi) \leq m$, then $\dim \operatorname{sing} u \leq m$.*

Remark: Of course the above corollary implies $\dim \operatorname{sing} u \leq m$ for *every* locally energy minimizing map $u \in W^{1,2}(\Omega; N)$ if N happens to be such that all homogeneous degree zero locally energy minimizing maps $\varphi \in W^{1,2}(\mathbb{R}^n; N)$ satisfy $\dim S(\varphi) \leq m$.

For example if $\dim N = 2$ and N has genus $g \geq 1$, we claim that this holds with $m = n-4$ (i.e. that $\dim S(\varphi) \leq n-4$ for every homogeneous degree zero locally energy

3.4. Further Properties of sing u

minimizing map from $\mathbb{R}^n$ into N), and hence by the corollary we have automatically sing $u = \mathcal{S}_{n-4}$. Indeed suppose there is a homogeneous degree zero locally energy minimizing map φ of $\mathbb{R}^n \to N$ with dim $S(\varphi) = n - 3$. Without loss of generality, we can assume that $S(\varphi) = \{0\} \times \mathbb{R}^{n-3}$, and we then have $\varphi(x, y) \equiv \varphi_0(|x|^{-1} x)$ for $x \in \mathbb{R}^3 \setminus \{0\}$ and $y \in \mathbb{R}^{n-3}$. But then φ_0 is a smooth non-constant minimizing map from S^2 into N. But such maps are known not to exist (see e.g. [Jo84]).

Proof of Corollary 1: By (ii), sing $u = \mathcal{S}_{n-3}$, hence the lemma with $j = n - 3$ gives precisely dim sing $u \leq n - 3$ as claimed. In case dim $N = 2$ and the genus of N is ≥ 1, we know by the above remark that sing $u = \mathcal{S}_{n-4}$, so in this case Lemma 1 gives dim sing $u \leq n - 4$.

Finally, if dim $S(\varphi) \leq m$ for all tangent maps φ, then by definition $\mathcal{S}_m = $ sing u and hence Lemma 1 gives dim sing $u \leq m$.

Proof of Lemma 1: We first prove that $\mathcal{S}_0 \cap \{x : \Theta_u(x) = \alpha\}$ is a discrete set for each $\alpha > 0$. Suppose this fails for some $\alpha > 0$. Then there are y, $y_j \in \mathcal{S}_0 \cap \{x : \Theta_u(x) = \alpha\}$ such that $y_j \neq y$ for each j, and $y_j \to y$. Let $\rho_j = |y_j - y|$ and consider the scaled maps u_{y, ρ_j}. By the discussion of Section 3.1 there is a subsequence $\rho_{j'}$ such that $u_{y, \rho_{j'}} \to \varphi$, where φ is (by definition) a tangent map of u at y; also, by Section 3.2 we have $\Theta_\varphi(0) = \Theta_u(y) = \alpha$.

Let $\xi_j = |y_j - y|^{-1}(y_j - y) (\in S^{n-1})$. We can suppose that the subsequence j' is such that $\xi_{j'}$ converges to some $\xi \in S^{n-1}$. Also (since the transformation $x \mapsto y + \rho_j x$ takes y_j to ξ_j) $\Theta_u(y_j) = \Theta_{u_{y, \rho_j}}(\xi_j) = \alpha$ for each j, hence by the upper semi-continuity of the density (as in Section 2.5) we have $\Theta_\varphi(\xi) \geq \alpha$. Thus since $\Theta_\varphi(x)$ has maximum value at 0 (by 3.3(i)), we have $\Theta_\varphi(\xi) = \Theta_\varphi(0) = \alpha$, and hence $\xi \in S(\varphi)$, contradicting the fact that $S(\varphi) = \{0\}$ by virtue of the assumption that $y \in \mathcal{S}_0$. □

Before we give the proof of the fact that dim $\mathcal{S}_j \leq j$, we need a preliminary lemma, which is of some independent interest. In this lemma, and subsequently, we use $\eta_{y, \rho}$ to be the map of $\mathbb{R}^n$ which translates y to the origin and homotheties by the factor ρ^{-1}; thus
$$\eta_{y, \rho}(x) = \rho^{-1}(x - y).$$

Lemma 2 *For each $y \in \mathcal{S}_j$, and each $\delta > 0$ there is an $\varepsilon > 0$ (depending on u, y, δ) such that for each $\rho \in (0, \varepsilon]$*
$$\eta_{y, \rho}\{x \in B_\rho(y) : \Theta_u(x) \geq \Theta_u(y) - \varepsilon\} \subset \text{ the } \delta\text{-neighbourhood of } L_{y, \rho}$$
for some j-dimensional subspace $L_{y, \rho}$ of $\mathbb{R}^n$ (see Figure 3.1).

Caution: We only prove this with the subspace $L_{y, \rho}$ depending on both y, ρ; thus, as ρ varies, even if y is fixed, the subspace $L_{y, \rho}$ may vary. See the example in Section 3.9 below.

Proof: If this is false, then there exists $\delta > 0$ and $y \in \mathcal{S}_j$ and sequences $\rho_k \downarrow 0$, $\varepsilon_k \downarrow 0$ such that
(1) $\quad \{x \in B_1(0) : \Theta_{u_{y, \rho_k}}(x) \geq \Theta_u(y) - \varepsilon_k\} \not\subset$ the δ-neighbourhood of L

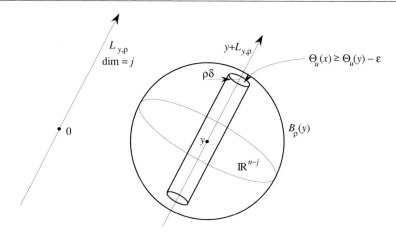

Figure 3.1: *Lemma 2*

for every j-dimensional subspace L of $\mathbb{R}^n$. But $u_{y,\rho_{k'}} \to \varphi$, a tangent map of u at y, and $\Theta_u(y) = \Theta_\varphi(0)$. Since $y \in \mathcal{S}_j$, we have $\dim S(\varphi) \leq j$, so (since $S(\varphi)$ is the set of points where Θ_φ takes its maximum value $\Theta_\varphi(0)$), there is a j-dimensional subspace $L_0 \supset S(\varphi)$ ($L_0 = S(\varphi)$ in case $\dim S(\varphi) = j$) and an $\alpha > 0$ such that

(2) $\qquad \Theta_\varphi(x) < \Theta_\varphi(0) - \alpha \quad$ for all $x \in \overline{B}_1(0)$ with $\mathrm{dist}\{x, L_0\} \geq \delta$.

Then we must have, for all sufficiently large k', that

(3) $\qquad \{x \in B_1(0) : \Theta_{u_{y,\rho_{k'}}}(x) \geq \Theta_\varphi(0) - \alpha\} \subset \{x : \mathrm{dist}\{x, L_0\} < \delta\}$.

Because otherwise we would have a subsequence $\{\tilde{k}\} \subset \{k'\}$ with $\Theta_{u_{y,\rho_{\tilde{k}}}}(x_{\tilde{k}}) \geq \Theta_\varphi(0) - \alpha$ for some sequence $x_{\tilde{k}} \in B_1(0)$ with $\mathrm{dist}\{x_{\tilde{k}}, L_0\} \geq \delta$. Taking another subsequence if necessary and using the upper semi-continuity result of Section 2.11, we get $x_{\tilde{k}} \to x$ with $\Theta_\varphi(x) \geq \Theta_\varphi(0) - \alpha$ and $\mathrm{dist}(x, L_0) \geq \delta$, contradicting (2).

Thus (3) is established, thus contradicting (1) for sufficiently large k. $\qquad\square$

Completion of the proof of Lemma 1: Define $\mathcal{S}_{j,i}$, $i \in \{1, 2, \dots\}$, defined to be the set of points y in $\mathcal{S}_j$ such that the conclusion of Lemma 2 above holds with $\varepsilon = i^{-1}$. Then, by Lemma 2, $\mathcal{S}_j = \cup_{i \geq 1} \mathcal{S}_{j,i}$. Next, for each integer $q \geq 1$ we let

$$\mathcal{S}_{j,i,q} = \{x \in \mathcal{S}_{j,i} : \Theta_u(x) \in (\tfrac{q-1}{i}, \tfrac{q}{i}]\},$$

and note that $\mathcal{S}_j = \cup_{i,q} \mathcal{S}_{j,i,q}$. For any $y \in \mathcal{S}_{j,i,q}$ we have trivially that

$$\mathcal{S}_{j,i,q} \subset \{x : \Theta_u(x) > \Theta_u(y) - \tfrac{1}{i}\},$$

and hence, by Lemma 2 (with $\varepsilon = i^{-1}$), for each $\rho \leq i^{-1}$

$$\eta_{y,\rho}(\mathcal{S}_{j,i,q} \cap B_\rho(y)) \subset \text{ the } \delta\text{-neighbourhood of } L_{y,\rho}$$

3.4. Further Properties of sing u

for some j-dimensional subspace $L_{y,\rho}$ of $\mathbb{R}^n$.

Thus each of the sets $A = \mathcal{S}_{j,i,q}$ has the "δ-approximation property" that there is ρ_0 $(= i^{-1}$ in the present case) such that, for each $y \in A$ and for each $\rho \in (0, \rho_0)$,

(*) $\qquad \eta_{y,\rho}(A \cap B_\rho(y)) \subset$ the δ-neighbourhood of $L_{y,\rho}$

for some j-dimensional subspace $L_{y,\rho}$ of $\mathbb{R}^n$.

In view of the arbitrariness of δ the proof is now completed by virtue of the following lemma:

Lemma 3 *There is a function $\beta : (0, \infty) \to (0, \infty)$ with $\lim_{t \downarrow 0} \beta(t) = 0$ such that if $\delta > 0$ and if A is an arbitrary subset of $\mathbb{R}^n$ having the property (*) above, then $\mathcal{H}^{j+\beta(\delta)}(A) = 0$.*

Proof: If $\delta \geq 1/8$ we can take $\beta(\delta) = n - j + 1$, so that $\mathcal{H}^{j+\beta(\delta)}(A) = \mathcal{H}^{n+1}(A) = 0$. Hence we can assume for the rest of the proof that $\delta \in (0, 1/8)$. First note that there is a fixed constant C_n such that for each $\sigma \in (0, 1/2)$ we can cover the closed unit ball $\overline{B}_1(0)$ of $\mathbb{R}^j$ with a finite collection of balls $\{B_\sigma(y_k)\}_{k=1,\ldots,Q}$ in $\mathbb{R}^j$ where $Q = Q(j)$ and $y_k \in \overline{B}_1(0)$, such that $Q\sigma^j < C_n$, and also $Q\sigma^{j+\beta} \leq 1/2$ for suitable $\beta = \beta(\sigma)$ with $\beta(\sigma) \downarrow 0$ as $\sigma \downarrow 0$.

It evidently follows that if L is any j-dimensional subspace of $\mathbb{R}^n$ and $\delta \in (0, 1/8)$, there is $\beta(\delta)$ (depending only on n, δ), with $\beta(\delta) \downarrow 0$ as $\delta \downarrow 0$, such that the 2δ-neighbourhood of $L \cap \overline{B}_1(0)$ can be covered by balls $B_\sigma(y_k)$, $k = 1, \ldots, \widetilde{Q}$, with $\sigma = 4\delta$ and with centers y_k in $L \cap B_1(0)$ and with $\widetilde{Q}\sigma^{j+\beta(\delta)} < \frac{1}{2}$. By scaling this means that for each $R > 0$ a $2\delta R$-neighbourhood of $L \cap B_R(0)$ can be covered by balls $B_{\sigma R}(y_k)$ with centers $y_k \in L \cap \overline{B}_R(0)$, $k = 1, \ldots, \widetilde{Q}$, such that $\widetilde{Q}(\sigma R)^{j+\beta(\delta)} < \frac{1}{2} R^{j+\beta(\delta)}$. The above lemma follows easily from this general fact by using successively finer covers of A by balls. The details are as follows: Supposing without loss of generality that A is bounded, we first take an initial cover of A by balls $B_{\rho_0/2}(y_k)$ with $A \cap B_{\rho_0/2}(y_k) \neq \emptyset$, $k = 1, \ldots, Q$, and let $T_0 = Q(\rho_0/2)^{j+\beta(\delta)}$. For each k pick $z_k \in A \cap B_{\rho_0/2}(y_k)$. Then by (*) with $\rho = \rho_0$ there is a j-dimensional affine space L_k such that $A \cap B_{\rho_0}(z_k)$ is contained in the δ-neighbourhood of L_k. Notice that $L_k \cap B_{\rho_0/2}(y_k)$ is a j-disk of radius $\leq \rho_0/2$, and so by the above discussion its $2\delta\rho_0$-neighbourhood (and hence also $A \cap B_{\rho_0/2}(z_k)$) can be covered by balls $B_{\sigma \rho_0/2}(z_{j,\ell})$, $\ell = 1, \ldots, P$, (centers not necessarily in A) such that $P(\sigma\rho_0/2)^{j+\beta(\delta)} \leq \frac{1}{2}(\rho_0/2)^{j+\beta(\delta)}$. Thus A can be covered by balls $B_{\sigma\rho_0/2}(w_\ell)$, $k = 1, \ldots, M$, such that $M(\sigma\rho_0/2)^{j+\beta(\delta)} \leq \frac{1}{2}T_0$. Proceeding iteratively we can thus for each q find a cover by balls $B_{\sigma^q \rho_0/2}(w_k)$, $k = 1, \ldots, R_q$, such that $R_q(\sigma^q \rho_0/2)^{j+\beta(\delta)} \leq 2^{-q}T_0$.

$\square$

3.5 Definition of Top-dimensional Part of the Singular Set

In this section we define the concept of "top dimensional part" of the singular set of u; actually we only consider the case here when this is $(n-3)$-dimensional (the generic case), but the reader should keep in mind that all the discussion here carries over with an integer $m \leq n-4$ in place of $n-3$ if the target manifold N happens to be such that all homogeneous degree zero maps $\varphi \in W^{1,2}_{\text{loc}}(\mathbb{R}^n; N)$ of u satisfy $\dim S(\varphi) \leq m$. (Recall that by the remark following the Corollary 1 in Section 3.4 this implies $\dim \operatorname{sing} u \leq m$ and in particular this is the case with $m = n-4$ when $\dim N = 2$ and N has genus ≥ 1.)

Definition 1 *The top dimensional part* $\operatorname{sing}_* u$ *of* $\operatorname{sing} u$ *is the set of points* $y \in \operatorname{sing} u$ *such that some tangent map* φ *of* u *at* y *has* $\dim S(\varphi) = n-3$.

Notice that then by definition we have $\operatorname{sing} u \setminus \operatorname{sing}_* u \subset \mathcal{S}_{n-4}$, and hence by Lemma 1 of Section 3.4 we have

(i) $$\dim(\operatorname{sing} u \setminus \operatorname{sing}_* u) \leq n-4.$$

To study $\operatorname{sing}_* u$ further, we first examine the properties of homogeneous degree zero minimizers $\varphi : \mathbb{R}^n \to N$ with $\dim S(\varphi) = n-3$.

3.6 Homogeneous Degree Zero φ with $\dim S(\varphi) = n-3$

Let $\varphi : \mathbb{R}^n \to N$ be any homogeneous degree zero minimizer with $\dim S(\varphi) = n-3$. Then, modulo an orthogonal transformation of $\mathbb{R}^n$ which takes $S(\varphi)$ to $\{0\} \times \mathbb{R}^{n-3}$, we have

(i) $$\varphi(x, y) \equiv \varphi_0(x),$$

where (x, y) denotes a general point in $\mathbb{R}^n$ with $x \in \mathbb{R}^3$, $y \in \mathbb{R}^{n-3}$, and where φ_0 is a homogeneous degree zero map from $\mathbb{R}^3$ into N. We in fact claim that

(ii) $$\operatorname{sing} \varphi_0 = \{0\} \quad \text{and hence } \varphi_0|S^2 \in C^\infty,$$

so that $\varphi_0|S^2$ is a smooth harmonic map of S^2 into N. To see this, first note that $\operatorname{sing} \varphi_0 \supset \{0\}$, otherwise φ_0, and hence φ, would be constant, thus contradicting the hypothesis $\dim S(\varphi) = n-3$. On the other hand if $\xi \neq 0$ with $\xi \in \operatorname{sing} \varphi_0$, then by homogeneity of φ_0 we would have $\{\lambda \xi : \lambda > 0\} \subset \operatorname{sing} \varphi_0$, and hence

$$\{(\lambda \xi, y) : \lambda > 0,\ y \in \mathbb{R}^{n-3}\} \subset \operatorname{sing} \varphi.$$

But the left side here is a half-space of dimension $(n-2)$, and hence this would give $\mathcal{H}^{n-2}(\operatorname{sing} \varphi) = \infty$, thus contradicting the fact that $\mathcal{H}^{n-2}(\operatorname{sing} \varphi) = 0$ by Lemma 1 of Section 2.10. Thus 3.6(ii) is established.

We also note that if $\varphi^{(j)}$ is any sequence of homogeneous degree zero minimizers with $\varphi^{(j)}(x,y) \equiv \varphi_0^{(j)}(x)$ for each j, and if $\limsup_{j\to\infty} \int_{B_1(0)} |D\varphi^{(j)}|^2 < \infty$, then

$$\limsup_{j\to\infty} \sup_{S^2} |D^\ell \varphi_0^{(j)}| < \infty$$

for each $\ell \geq 0$. Indeed by the compactness theorem (Lemma 1 of Section 2.9) there is a subsequence $\varphi^{(j')} \to \varphi$, where φ is a homogeneous degree zero minimizer with $\varphi(x,y) = \varphi_0(x)$, and by (ii) we have $\varphi_0|S^2 \in C^\infty$. Thus for $z \in \mathbb{R}^n \setminus (\{0\} \times \mathbb{R}^{n-3})$ and positive $\sigma < \operatorname{dist}(z, \{0\} \times \mathbb{R}^{n-3})$ we have $|\varphi(x) - \varphi(z)| \leq C\sigma$, whence $\int_{B_\sigma(z)} |\varphi^{(j')} - \varphi(z)|^2 \leq \int_{B_\sigma(z)} |\varphi^{(j')} - \varphi|^2 + C\sigma^2$, so we can apply the regularity theorem of Section 2.3 for σ sufficiently small and j' sufficiently large, so that the convergence of $\varphi^{(j')}$ to φ is actually with respect the C^k norm for each k on compact subsets of $\mathbb{R}^n \setminus (\{0\} \times \mathbb{R}^{n-3})$. In view of the arbitrariness of the sequence $\varphi^{(j')}$, it then follows that for all homogeneous degree zero minimizers φ with $\int_{S^2} |D\varphi_0|^2 \leq \Lambda$ we have

(iii) $$\sup_{S^2} |D^\ell \varphi_0| \leq C, \quad \ell = 1, 2, \ldots,$$

where C depends only on ℓ, N, Λ.

3.7 The Geometric Picture Near Points of sing$_*u$

Let K be a compact subset of Ω and $z \in \operatorname{sing}_* u \cap K$, and let φ be a tangent map of u at z with $\dim S(\varphi) = n-3$. As in Section 3.4, we can assume without loss of generality (after making an orthogonal transformation in $\mathbb{R}^n$ which takes $S(\varphi)$ to $\{0\} \times \mathbb{R}^{n-3}$), that

(i) $$\varphi(x,y) \equiv \varphi_0(x), \quad x \in \mathbb{R}^3, y \in \mathbb{R}^{n-3}.$$

By definition of $\operatorname{sing}_* u$, there is a sequence $\rho_j \downarrow 0$ such that

(ii) $$\lim_{j\to\infty} \rho_j^{-n} \int_{B_{\rho_j}(z)} |u - \varphi^{(z)}|^2 = 0,$$

where $\varphi^{(z)}(x,y) \equiv \varphi((x,y) - z)$ so for $\rho = \rho_j$ with j sufficiently large we can make the scaled L^2-norm $\rho^{-n} \int_{B_\rho(z)} |u - \varphi^{(z)}|^2$ as small as we wish. On the other hand we claim that for any homogeneous degree zero minimizing maps $\varphi : \mathbb{R}^n \to N$ as in 3.6(i) and any ball $B_{\rho_0}(z)$ with $\overline{B}_{\rho_0}(z) \subset \Omega$ we have the estimate

(iii) $$\operatorname{sing} u \cap B_{\rho/2}(z) \subset \{x : \operatorname{dist}(x, (z + \{0\} \times \mathbb{R}^{n-3})) < \delta(\rho)\rho\} \quad \forall \rho \leq \rho_0,$$

$$\delta(\rho) = C\left(\rho^{-n} \int_{B_\rho(z)} |u - \varphi^{(z)}|^2\right)^{\frac{1}{n}},$$

where C depends only on n, N, Λ with Λ any upper bound for $\rho_0^{2-n}\int_{B_{\rho_0}(z)}|Du|^2$ (see Figure 3.2).

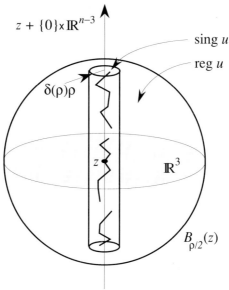

Figure 3.2: *Picture near points of* $\text{sing}_* u$

In view of 3.3(ii), this perhaps suggests that the possibility that the top dimensional part of the singular set is contained in a C^1 manifold (or at least a Lipschitz manifold) of dimension $n-3$. But there is a problem in that 3.3(ii) only guarantees that $\delta(\rho)$ is small when ρ is proportionally close to one of the ρ_j, and, without further input, we cannot conclude very much about the structure of $\text{sing}_* u$ from this—see the discussion in Section 3.8 below.

We conclude this section with the simple proof of (iii). We assume $z=0$ for convenience of notation.

Proof: Let $\rho < \rho_0$ and $w = (\xi, \eta) \in \text{sing}_* u \cap B_{\rho/2}(0)$. Take $\sigma = \beta_0|\xi|$, with $\beta_0 \leq \frac{1}{2}$ to be chosen. By the regularity theorem Section 2.3 there is $\varepsilon_0 = \varepsilon_0(n, N, \Lambda) > 0$ such that

$$(1) \qquad \varepsilon_0 \leq \sigma^{-n}\int_{B_\sigma(w)}|u-\varphi(w)|^2 \leq$$

$$\leq 2\sigma^{-n}\int_{B_\sigma(w)}|u-\varphi|^2 + 2\sigma^{-n}\int_{B_\sigma(w)}|\varphi-\varphi(w)|^2.$$

By virtue of 3.6(iii) we know that $|D\varphi_0(x)| \leq C|x|^{-1}$, where C depends only on N, Λ, and hence $|\varphi(w) - \varphi(x)| \leq C|\xi|^{-1}\sigma \leq C\beta_0$ for $x \in B_\sigma(w)$, where C depends only on N and Λ. Then (1) gives

$$\varepsilon_0 \leq 2\beta_0^{-n}|\xi|^{-n}\int_{B_\rho(0)}|u-\varphi|^2 + C\beta_0^2.$$

Then selecting $C\beta_0^2 \leq \frac{1}{2}\varepsilon_0$ and multiplying through by $|\xi|^n$, we have

$$|\xi|^n \leq C\left(\rho^{-n}\int_{B_\rho(0)}|u-\varphi|^2\right)\rho^n,$$

with C depending only on n, N, Λ. Taking n^{th} roots of each side, we then get the required inequality. □

3.8 Consequences of Uniqueness of Tangent Maps

We want to show here that the geometric picture established in the previous section does give good information about the structure of the top dimensional part $\text{sing}_* u$ if the tangent map at each point $y \in \text{sing}_* u$ is *unique*; because then for each $\delta > 0$ and each $y \in \text{sing}_* u$ there is φ as above, an orthogonal transformation Q of $\mathbb{R}^n$ and a $\rho_{y,\delta} > 0$ such that 3.7(iii) holds for all $\rho \leq \rho_{y,\delta}$, with Q independent of ρ. We claim that such a property implies that $\text{sing}_* u$ is contained in a countable union of $(n-3)$-dimensional Lipschitz graphs: To be precise, we could apply the case $j = n - 3$ of the following lemma:

Lemma 1 *Let $j \in \{1, \ldots, n-1\}$. Suppose $\delta \in (0,1)$ and A is a subset of $\mathbb{R}^n$ such that at each point $y \in A$ there is a j-dimensional subspace L_y of $\mathbb{R}^n$ and $\rho_y > 0$ such that*

(i) $\qquad A \cap B_\rho(y) \subset \{x : \text{dist}(A \cap B_\rho(y), y + L_y) \leq \delta\rho\} \qquad \forall \rho \leq \rho_y,$

then $A \subset \cup_{i=1}^\infty \Sigma_i$, where each Σ_i is the graph of a Lipschitz function over some j-dimensional subspace (in the sense that there is an open subset U_i of some j-dimensional subspace $L_i \subset \mathbb{R}^n$ and an $L_i^\perp$-valued Lipschitz function f_i on an open subset U_i of L_i such that $\Sigma_i = \{x + f_i(x) : x \in U_i\}$).

Remark: In standard terminology, this says that A is countably j-rectifiable.

Proof: We decompose $A = \cup_{i=1}^\infty A_i$, where $A_{i+1} \subset A_i$ is the set of points $y \in A$ such that (i) holds with $\rho_y = i^{-1}$. Notice that then (i) holds for all $y \in A_i$ with $\rho_y \equiv i^{-1}$, and A_i satisfies a uniform cone condition—in the sense that

$$A_i \cap B_{i^{-1}}(y) \subset K_y, \qquad \forall y \in A_i,$$

where K_y is the cone given by $K_y = \{x : \text{dist}(x, y + L_y) < \delta|x - y|\}$. Now select j-dimensional subspaces $L_1, \ldots, L_Q$ of $\mathbb{R}^n$ such that for each j-dimensional subspace $L \subset \mathbb{R}^n$ there is one of the L_j such that $\|L_j - L\| < \delta$. Then we can decompose $A_i = \cup_{j=1}^Q A_{i,j}$, where

$$A_{i,j} = \{y \in A_i : \|L_y - L_j\| < \delta\}.$$

Then each $A_{i,j}$ has the uniform cone property that

$$A_{i,j} \cap B_{i-1}(y) \subset y + K_j, \qquad \forall\, y \in A_{i,j},$$

where $K_j = \{x : \operatorname{dist}(x, L_j) < 2\delta|x|\}$. It is standard that such a uniform cone condition implies that, for each given $y \in A_{i,j}$, $A_{i,j} \cap B_{i-1}(y)$ is contained the graph of a Lipschitz function with domain $B_{i-1}(y') \cap L_j$, where y' is the orthogonal projection of y on L_j. (See e.g. [Si83a, §5].) The lemma is thus proved. □

Notice that the argument above actually shows that the following is true:

Corollary 1 *If A satisfies the same hypotheses as in the lemma, except that we can choose $\rho_y \equiv \rho_0$, with $\rho_0 > 0$ independent of y, then $A \cap B_{\rho_0}(y)$ is contained in the finite union of j-dimensional Lipschitz graphs for each $y \in A$. If $L_y - y$ can be selected independent of y (i.e. $L_y = y + L_0$ for some fixed subspace L_0), then $A \cap B_{\rho_0}(y)$ is contained in the graph of a Lipschitz function over the plane L_0.*

By 3.7(iii), such a uniform choice of ρ_y, L_y can be made for $\operatorname{sing}_* u \cap K$, K any compact subset of Ω, provided that for each $\delta > 0$ and each compact $K \subset \Omega$, there is $\rho(\delta, K) \in (0, \operatorname{dist}(K, \partial\Omega))$ such that

(ii) $\qquad \rho^{-n} \int_{B_\rho(y)} |u - \varphi|^2 < \delta, \quad \rho < \rho(\delta, K),\ y \in \operatorname{sing}_* u \cap K.$

As a matter of fact by 3.7(iii) we only need (ii) to hold for suitable $\delta = \delta(n, N, K, \Lambda) > 0$, where Λ is any upper bound for $\sup d^{2-n} \int_{B_d(y)} |Du|^2$ over all $y \in K$ with $d \in (0, \operatorname{dist}(K, \partial\Omega))$, because then 3.7(iii) implies that the hypotheses of the above corollary with $\delta =$ (for example) $\frac{1}{2}$. Thus:

Corollary 2 *There is $\delta = \delta(n, N, K, \Lambda) > 0$ such that if for each $y \in \operatorname{sing}_* u \cap K$ there is φ such that (ii) holds, then $\operatorname{sing}_* u \cap B_{\rho(\delta, K)}(y)$ is contained in an $(n-3)$-dimensional Lipschitz graph for each $y \in \operatorname{sing}_* u \cap K$.*

We see in the next chapter that there are stronger conditions on the L^2-norm which guarantee much stronger results in certain cases.

3.9 Approximation properties of subsets of $\mathbb{R}^n$

We want to devote this section to some further discussion of properties of subsets $A \subset \mathbb{R}^n$ which satisfy the kind of j-dimensional approximation property described in Lemma 1 of the previous section.

We in fact consider several variants of such a property; we continue to use the notation that

$$\eta_{y,\rho}(x) = \rho^{-1}(x - y).$$

3.9. Approximation properties of subsets of $\mathbb{R}^n$

Definition 1 *Let $A \subset \mathbb{R}^n$ be an arbitrary set and $\delta > 0$; then*

(i) A has the weak *j-dimensional δ-approximation property if $\forall\, y \in A$ there is $\rho_y > 0$ such that, $\forall\, \rho \in (0, \rho_y]$, $B_1(0) \cap \eta_{y,\rho}(A) \subset$ the δ-neighbourhood of some j-dimensional affine space $L_{y,\rho}$ containing y.*

(ii) The property in (i) is said to be ρ_0-uniform, if A is contained in some ball of radius ρ_0 and if, for every $y \in A$ and every $\rho \in (0, \rho_0]$, $B_1(0) \cap \eta_{y,\rho}(A) \subset$ the δ-neighbourhood of some j-dimensional affine space $L_{y,\rho}$ containing y.

(iii) A has the strong *j-dimensional δ-approximation property if for each $y \in A$ there is a j-dimensional affine space L_y containing y such that definition (i) holds with $L_{y,\rho} = L_y$ for every $\rho \in (0, \rho_y]$.*

(iv) The property in (iii) is said to be ρ_0-uniform if A is contained in some ball of radius ρ_0 and if for each $y \in A$ there is a j-dimensional affine space L_y containing y such that $B_1(0) \cap \eta_{y,\rho}(A) \subset$ the δ-neighbourhood of L_y for each $\rho \in (0, \rho_0]$.

Concerning these properties, we have the following lemma, which is actually just a summary of the results of Lemma 3, Lemma 1, and Corollary 1 of Section 3.8. We continue to use the terminology that G *is the graph of a Lipschitz function over some j-dimensional subspace* to mean that there is a j-dimensional subspace $L \subset \mathbb{R}^n$ and a map $u: L \to L^\perp$ such that $\sup_{x,y \in L, x \neq y} |x-y|^{-1}|u(x) - u(y)| < \infty$ and $G = \{x + u(x) : x \in L\}$.

Lemma 1 *(i) There is a function $\beta : [0, \infty) \to [0, \infty)$ with $\lim_{\delta \downarrow 0} \beta(\delta) = 0$ such that if $A \subset \mathbb{R}^n$ has the j-dimensional weak δ-approximation property for some given $\delta \in (0, 1]$, then $\mathcal{H}^{j+\beta(\delta)}(A) = 0$. (In particular, if A has the j-dimensional weak δ-approximation property for each $\delta > 0$, then $\dim A \leq j$.)*

(ii) If $A \subset \mathbb{R}^n$ has the strong j-dimensional δ-approximation property for some $\delta \in (0, 1]$, then $A \subset \cup_{k=1}^\infty G_k$, where each G_k is the graph of some Lipschitz function over some j-dimensional subspace of $\mathbb{R}^n$.

(iii) If $A \subset \mathbb{R}^n$ has the ρ_0-uniform strong j-dimensional δ-approximation property for some $\delta \in (0, 1]$, then $A \subset \cup_{k=1}^Q G_k$, where each G_k is the graph of some Lipschitz function over some j-dimensional subspace of $\mathbb{R}^n$.

The following lemma shows that certain closed subsets of the singular set of u satisfy the uniform weak δ-approximation property:

Lemma 2 *If $u \in W^{1,2}(\Omega, N)$ is energy minimizing, if $y_0 \in \operatorname{sing} u$, if $S_+ = \{x \in \Omega : \Theta_u(y) \geq \Theta_u(y_0)\}$, and if $\delta > 0$ is arbitrary, then $S_+ \cap B_{\rho_0}(y_0)$ has the $(n-3)$-dimensional ρ_0-uniform weak δ-approximation property for suitable $\rho_0 = \rho_0(u, y_0, \delta) > 0$.*

Remarks: $S_+ \subset \operatorname{sing} u$ because $\Theta_u(y) > 0 \iff y \in \operatorname{sing} u$ by the ε-regularity theorem—see Corollary 2 of Section 2.10. Notice also that S_+ is a *closed* subset of $\operatorname{sing} u$ by the upper semi-continuity of Θ proved in Section 2.5.

Proof: If the lemma is false, then there is $\delta > 0$, $y \in \operatorname{sing} u$, $\rho_k \downarrow 0$, $\sigma_k < \rho_k$, and $y_k \in B_{\rho_k}(y) \cap S_+$ such that

(1) $$B_1(0) \cap \eta_{y_k,\sigma_k} S_+ \not\subset \delta\text{-neighbourhood of any } (n-3)\text{-dimensional subspace}.$$

Choose $R_k \downarrow 0$ with $R_k/\rho_k \to \infty$. Then by monotonicity (see Section 2.4) we have, for all $\rho \in (0, R_k]$ and for all $k = 1, 2, \ldots$,

$$\Theta_u(y_k) \leq \rho^{2-n} \int_{B_\rho(y_k)} |Du|^2 \leq R_k^{2-n} \int_{B_{R_k}(y_k)} |Du|^2 \leq R_k^{2-n} \int_{B_{R_k+\rho_k}(y_0)} |Du|^2.$$

In terms of the re-scaled function $u_k = u_{y_k,\sigma_k}$ this says

$$\Theta_u(y_k) \leq \rho^{2-n} \int_{B_\rho(0)} |Du_k|^2 \leq R_k^{2-n} \int_{B_{R_k+\rho_k}(y_0)} |Du|^2$$

for every $\rho > 0$ and for all sufficiently large k (depending on ρ). Since $\rho_k/R_k \to 0$ we have $R_k^{2-n} \int_{B_{R_k+\rho_k}(y_0)} |Du|^2 \to \Theta_u(y_0)$, and since $\Theta_u(y_k) \geq \Theta_u(y)$ by hypothesis, we then obtain

(2) $$\Theta_u(y_0) \leq \rho^{2-n} \int_{B_\rho(0)} |Du_k|^2 \leq \Theta_u(y_0) + \varepsilon_k,$$

where $\varepsilon_k \to 0$ as $k \to \infty$. In particular the u_k have uniformly bounded energy on any fixed ball in $\mathbb{R}^n$, so the compactness theorem (Lemma 1 of Section 2.9) gives that there is a minimizing map $\varphi \in W^{1,2}_{\text{loc}}(\mathbb{R}^n; N)$ and a subsequence $u_{k'}$ such that $u_{k'} \to \varphi$ locally on $\mathbb{R}^n$ both with respect to L^2 and with respect to energy. But then (2) says that

$$\rho^{2-n} \int_{B_\rho(0)} |D\varphi|^2 \equiv \Theta_u(y_0), \quad \forall \rho > 0,$$

and by the monotonicity formula (Section 2.4) applied to φ we thus conclude that

$$\int_{B_\rho(0)} R^{2-n} \left|\frac{\partial \varphi}{\partial R}\right|^2 = 0, \quad \forall \rho > 0,$$

and hence that φ is homogeneous of degree zero:

(3) $$\varphi(\lambda x) \equiv \varphi(x), \quad x \in \mathbb{R}^n, \lambda > 0.$$

Henceforth let $\alpha = \Theta_u(y_0) (\equiv \Theta_\varphi(0))$. Now φ need not be a tangent map of u because the points y_k vary with k, but in any case we can (by (3)) apply the discussion of Section 3.3 in order to deduce that

$$S(\varphi) := \{y \in \mathbb{R}^n : \Theta_\varphi(y) = \alpha\}$$

3.9. Approximation properties of subsets of $\mathbb{R}^n$

is contained in an $(n-3)$-dimensional subspace L of $\mathbb{R}^n$. Thus by upper semi-continuity of Θ_φ there is $\theta > 0$ such that

(4) $$\{y \in \overline{B}_1(0) : \Theta_\varphi(y) \geq \alpha - \theta\} \subset L_\delta,$$

where L_δ denotes the (open) δ-neighbourhood of L. Now by the upper semi-continuity of Θ (as in Section 2.11) we see immediately that this implies

(5) $$\{y \in \overline{B}_1(0) : \Theta_{u_{k'}}(y) \geq \alpha - \theta\} \subset L_\delta$$

for all sufficiently large k'. Indeed otherwise there would be a subsequence $\{\tilde{k}\} \subset \{k'\}$ and $x_{\tilde{k}} \in \overline{B}_1(0) \setminus L_\delta \to x \in \overline{B}_1(0) \setminus L_\delta$ and with $\Theta_{u_{\tilde{k}}}(x_{\tilde{k}}) \geq \alpha - \theta$. But then by the upper semi-continuity we have $\Theta_\varphi(x) \geq \alpha - \theta$ with $x \in \overline{B}_1(0) \setminus L_\delta$, which contradicts (4). Thus (5) is established. But evidently (5) contradicts (1). □

We want to conclude this section by briefly discussing an example which illustrates the point that "very bad" sets A may have the weak j-dimensional approximation property.

We begin with an isosceles triangle A_0 with edge-lengths $\ell, \ell, 1$, and with angles $\varepsilon, \varepsilon, \pi - 2\varepsilon$, where $\varepsilon \in (0, \pi/4)$ (see Figure 3.3); we should imagine ε small. We

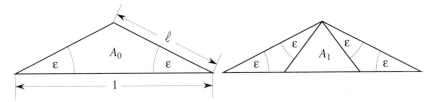

Figure 3.3: *The triangle A_0 and construction for A_1*

proceed to describe an iterative procedure which, at the kth stage gives rise to a union of 2^k homothetic copies of A_0, each scaled by a scaling factor ℓ^k. (Note that for small ε, $\ell = \frac{1}{2} + \frac{1}{4}\varepsilon^2 + O(\varepsilon^4)$.) Specifically, inductively define the sequence A_k, with $A_{k+1} \subset A_k$ as follows:

A_1 is defined as the union of the two homothetic copies of A_0 obtained by joining the base of A_0 to the vertex of A_0 opposite the base with line segments making angles ε with the two equal edges of A_0. (See Figure 3.3.) Thus both the triangles in A_1 has edge-lengths ℓ^2, ℓ^2, ℓ.

Then A_2 is the union of the four triangles (having edges of length ℓ^3, ℓ^3, ℓ^2) obtained by applying the same construction to each of the two triangles in A_1. At the kth step we apply the same construction to each of the 2^{k-1} triangles (with edge-lengths $\ell^k, \ell^k, \ell^{k-1}$) in A_{k-1}. Thus A_k consists of a union of 2^k triangles, each having edge-lengths $\ell^{k+1}, \ell^{k+1}, \ell^k$. Then we define

$$\Gamma = \cap_{k=1}^\infty A_k.$$

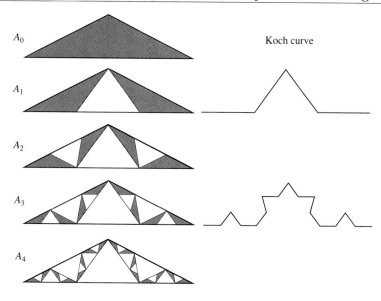

Figure 3.4: *The sets A_k and the Koch curve*

Of course Γ is a compact subset of A_k for each k and has full projection onto the base of A_0, because each A_k evidently has this property.

Notice that in fact Γ is just one of the well-known "Koch curves", because the union of the shorter edges of all the triangles in A_{2k+1} is just the standard k^{th} approximation for the Koch curve with starting figure consisting of 4 segments each of length ℓ as shown in Figure 3.4.

Since the Hausdorff dimension d of Γ satisfies $\left(\frac{1}{\ell}\right)^d = 2$ (we use two copies of Γ scaled by factor ℓ to reproduce Γ), we get $d = -\frac{\log 2}{\log \ell} \in (1,2)$ for $\varepsilon \in (0, \frac{\pi}{4})$. Γ is an example of a self-similar set; if we homothety by a factor of ℓ^{-1} from any vertex of any one of the triangles in the collection $\cup_{k\geq 1} A_k$ and intersect the result with a suitable ball with center at the chosen vertex, then we see a copy of Γ modulo a rigid motion. (See [Hu81] for a general discussion of such sets.) Using this self-similarity property, and the corresponding similarity properties of the approximations A_k of Γ it is quite easy to check that Γ has the (alarmingly bad) property that, even though $\mathcal{H}^1(\Gamma) = \infty$, nevertheless $\mathcal{H}^1(\gamma \cap \Gamma) = 0$ for any embedded C^1 curve γ or for any graph γ of a Lipschitz function defined over a 1-dimensional subspace of $\mathbb{R}^2$. (Such subsets are called "purely 1-unrectifiable": see [Hu81], [Mt75], [Mt82], [Mt84], [Ms53] or [Ha85].) On the other hand, again by using the self-similarity properties of A_k it is elementary to check that Γ has the weak 1-dimensional δ-approximation property with $\delta = 6\varepsilon$.

Thus there is are extremely significant differences between subsets which satisfy the weak and strong j-dimensional δ-approximation properties; for example Lemma 1 above shows that subsets with the strong j-dimensional δ-approximation property are automatically countably j-rectifiable, whereas the above example shows that sets with the weak j-dimensional approximation property can have Hausdorff dimension greater than j and be purely j-unrectifiable.

Thus although Lemma 2 will prove to be useful in our later discussion, it does not in itself guarantee very much about the structure of sing u.

3.10 Uniqueness of Tangent maps with isolated singularities

Recall that we in fact could check the strong $(n-3)$-dimensional δ-approximation property (and hence countable rectifiability, by Lemma 1) of the previous section for sing u (u any energy minimizing map $u \in W^{1,2}(\Omega; N)$), *provided* all the tangent maps of u at points of $\operatorname{sing}_* u$ are unique. Unfortunately such uniqueness in general is not true, although it is still an open question in case the target is real-analytic.

In this section we discuss one of the few situations in which uniqueness of tangent maps is known—the case when the tangent map has only an isolated singularity at 0 and when the target manifold N is real-analytic. The main theorem here (originally proved in [Si83b]) is as follows:

Theorem 1 *Suppose φ is a tangent map of u at some point $y \in \operatorname{sing} u$, and suppose $\operatorname{sing} \varphi = 0$. Also, assume that N is real-analytic. Then φ is the unique tangent map for u at y, and in fact*

$$u(y + r\omega) = \varphi(\omega) + \varepsilon(r, \omega), \qquad \omega \in S^{n-1},$$

where $\lim_{r \downarrow 0} |\log r|^\alpha \sup_{\omega \in S^{n-1}} |\varepsilon(r,\omega)| = 0$ for some $\alpha > 0$.

Remark: (1) In view of the examples constructed in [AS88] and [GW89] the decay here is best possible.

(2) The theorem is not true if we replace the hypothesis that N is real-analytic by the hypothesis that N is C^∞ (see [Wh92]). We briefly discuss why such a state of affairs, which might seem surprising at first reading, is to be expected. First, rewrite the equation $\Delta u + \sum_{j=1}^n A_u(D_j u, D_j u) = 0$ (see Section 2.2) in terms of spherical coordinates $r = |x-y|$, $\omega = |x-y|^{-1}(x-y)$, and then make the further change of variable $t = -\log r$. Letting $\widetilde{u}(\omega, t) = u(y + r\omega)$ and letting $\widetilde{u}'$ abbreviate $\partial \widetilde{u}/\partial t$, we obtain the equation

(1) $$\widetilde{u}'' - (n-2)\widetilde{u}' + \Delta_{S^{n-1}}\widetilde{u} + A_{\widetilde{u}}(D_\omega \widetilde{u}, D_\omega \widetilde{u}) = -A_{\widetilde{u}}(\widetilde{u}', \widetilde{u}'),$$

where $A_{\widetilde{u}}(D_\omega \widetilde{u}, D_\omega \widetilde{u})$ is an abbreviation for $\sum_{j=1}^{n-1} A_{\widetilde{u}}(D_{\tau_j}\widetilde{u}, D_{\tau_j}\widetilde{u})$, where $\tau_1, \ldots, \tau_{n-1}$ is any orthonormal basis for the tangent space of the sphere S^{n-1}.

Now by modifying the argument of Section 2.2 in a straightforward way, we can check that $\Delta_{S^{n-1}}\tilde{u} + A_{\tilde{u}}(D_\omega \tilde{u}, D_\omega \tilde{u})$ is exactly the Euler-Lagrange operator corresponding to the energy functional $\mathcal{E}_{S^{n-1}}(v) := \int_{S^{n-1}} |D_\omega v|^2$ for maps $v : S^{n-1} \to N$. Notice that then of course, since φ is homogeneous of degree zero and $\operatorname{sing}\varphi = \{0\}$ we know that $\varphi_0 := \varphi|S^{n-1} : S^{n-1} \to N$ is C^∞ and stationary for $\mathcal{E}_{S^{n-1}}$. Thus

(2) $$\Delta_{S^{n-1}}\varphi_0 + A_u(D_\omega \varphi_0, D_\omega \varphi_0) = 0.$$

Also, keep in mind that the Euler-Lagrange operator for $\mathcal{E}_{S^{n-1}}$ (i.e., the operator on the left of (2)) is by definition (see the discussion in Section 2.2) characterized as the operator $\mathcal{N}(\varphi_0)$ such that

$$\langle \mathcal{N}(\varphi_0), \zeta \rangle = -\frac{d}{ds}\mathcal{E}_{S^{n-1}}(\Pi(\varphi_0 + s\zeta))|_{s=0}, \quad \zeta = (\zeta^1, \ldots, \zeta^p) \in C^\infty(S^{n-1}; \mathbb{R}^p);$$

that is, it can be thought of as the L^2 gradient of the functional $\mathcal{E}_{S^{n-1}}$, and hence it is reasonable to use the alternative notation $-\nabla \mathcal{E}_{S^{n-1}}$ for this operator. In this case the equation (1) can be written

(3) $$\tilde{u}'' - (n-2)\tilde{u}' - \nabla \mathcal{E}_{S^{n-1}}(\tilde{u}) = -A_{\tilde{u}}(\tilde{u}', \tilde{u}').$$

Now since φ is a tangent map for u at y, we know there is some sequence $\rho_j \downarrow 0$ such that u_{y,ρ_j} converges to φ in the $W^{1,2}$ norm on any ball $B_R(0)$; here as usual $u_{y,\rho}$ denotes the re-scaled function given by $u_{y,\rho}(x) = u(y + \rho x)$. Since φ is homogeneous of degree zero and smooth away from 0, and hence $|D\varphi(x)|^2 \leq C|x|^{-2}$, this ensures in particular that, for any given $\theta \in (0,1)$, $\sigma^{2-n}\mathcal{E}_{B_\sigma(z)}(u_{y,\rho_j}) \leq C\theta^{-2}\sigma^2$, provided $z \in \mathbb{R}^n \setminus B_\theta(0)$, $\sigma \in (0, \theta/2)$ and j is sufficiently large (depending on θ). Then by the version of the ε-regularity theorem given in Corollary 1 in Section 2.10 we have that, for suitable $\varepsilon = \varepsilon(n, N) > 0$ and all sufficiently large j,

$$\sigma < \varepsilon\theta \implies \sup_{B_\sigma(z)} \sigma^k |D^k u_{y,\rho_j}| \leq C_k$$

for each $k = 0, 1, \ldots$ and for j sufficiently large (depending on θ, k); of course since any region $B_R(0) \setminus B_\rho(0)$ can be covered by finitely many balls $B_\sigma(z)$ with $\sigma \leq \varepsilon\theta$ this shows that in fact for any fixed $0 < \rho < R$

(4) $$\sup_{B_R(0) \setminus B_\rho(0)} |D^k u_{y,\rho_j}| \leq C(k, \rho, R)$$

for all $k = 0, 1, \ldots$ and for all sufficiently large j (depending on ρ, R, k). Then of course the convergence of u_{y,ρ_j} to φ must be with respect to the C^k norm on any compact subset of $\mathbb{R}^n$ and in particular, since $\frac{\partial \varphi}{\partial r} = 0$, we have

(5) $$\frac{\partial u_{y,\rho_j}}{\partial r} \to 0 \text{ on each compact subset of } \mathbb{R}^n \setminus \{0\}.$$

Thus in terms of the function $\tilde{u}(\omega, t)$, we see that with $T_j = -\log \rho_j \to \infty$, $\tilde{u}(\omega, t - T_j)$ converges to $\varphi(\omega)$ in the C^k norm on all compact subsets of the cylinder $S^{n-1} \times [0, \infty)$ and in particular $\tilde{u}'(\omega, t - T_j)$ converges to zero on any such set. Thus fixing

3.10. Uniqueness of Tangent maps with isolated singularities

$T \geq 4$, taking arbitrary j sufficiently large, and letting $v(\omega, t) = u(\omega, t - T_j)$, we see that the equation (3) can be written in the form

$$v'' - (n-2)v' = \nabla \mathcal{E}_{S^{n-1}}(v) + R, \quad \text{on } S^{n-1} \times [0, T)$$

where $\|R(t)\|_{L^2(S^{n-1})} \leq \frac{1}{2}\|v'(t)\|_{L^2(S^{n-1})}$.

Now (after a re-scaling of the time variable to get rid of the factor $(n-2)$ in front of v') this is analogous to a finite dimensional ODE system of the form

(6) $$\xi'' - \xi' = \nabla f(\xi) + R,$$

where f is a given fixed smooth function on $\mathbb{R}^n$ and $|R| \leq \frac{1}{2}|\xi'|$. Of particular interest are "slow decay solutions" of such an equation; that is, a solution ξ such that $|\xi''| \ll |\xi'|$. (Such solutions exist whenever f has critical points which are degenerate in a certain bad sense—see [AS88]). In this case the equation (6) takes the form

(7) $$\xi' = -\nabla f(\xi) + \tilde{R},$$

where $|\tilde{R}| \leq \frac{3}{4}|\xi'|$. Now the asymptotic behaviour of such equations is very different in the cases $f \in C^\infty$ and f real analytic. In fact, it is easy to construct examples of C^∞ functions f such that there are solutions ξ of an equation of the form (7) (or even exact solutions of the equations $\xi' = -\nabla f(\xi)$, $\xi'' - \xi' = \nabla f(\xi)$) which have no limit as $t \to \infty$, even though the solution in question remains in a compact region for all time. Indeed there are examples of C^∞ functions f with f and its gradient vanishing on a smooth Jordan curve γ, with f positive inside γ, and such that there are solutions of $\xi' = -\nabla f(\xi)$ (or $\xi'' - \xi' = \nabla f(\xi)$) which "spiral out" towards γ ("goat tracks down the hillside") as $t \to \infty$, so that the set of limit points of $\xi(t)$ as $t \to \infty$ is all of γ. See Figure 3.5.

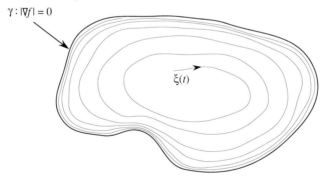

Figure 3.5: *"goat tracks" for C^∞-potential*

On the other hand if f is real-analytic the situation is very different: According to Lojasiewicz [Lo65], for each critical point y of f (i.e., each point where $\nabla f = 0$) there is $\alpha \in (0,1)$ and $\eta > 0$ such that

(i) $$|\nabla f(x)| \geq |f(x) - f(y)|^{1-\alpha/2}$$

for every point $x \in B_\eta(y)$. We emphasize that this holds for *any* f which is real analytic in some neighbourhood of 0; the constants α, η of course depend on the particular function f. Using this inequality it is easy to prove that any solution of an equation of the form (7) which remains in a compact region for all time must indeed have a unique asymptotic limit as $t \to \infty$:

Theorem 2 *Suppose $f \in C^\omega(\mathbb{R}^n)$ and $\xi : [0, \infty) \to \mathbb{R}^n$ is a bounded $C^1([0, \infty))$ solution of $\xi' = -\nabla f(\xi) + R$, where R has the property that there is a fixed $\theta \in (0, 1)$ such that $|R| \leq \theta|\xi'|$ on $[0, \infty)$. Then $\lim_{t \to \infty} \xi(t)$ exists, and in fact there are constants $C, \alpha > 0$ and $\xi_0 \in \mathbb{R}^n$ such that $|\xi(t) - \xi_0| \leq C(\log t)^{-\alpha}$ for all $t \geq 2$.*

Since our proof of Theorem 1 will involve an analogous argument in an infinite dimensional setting, it is worthwhile for us to give the proof:

Proof: First note that by direct computation we have

$$-\frac{d}{dt}f(\xi(t)) = -\xi' \cdot \nabla f(\xi) \equiv |\xi'|^2 - \xi' \cdot R \geq (1 - \theta)|\xi'|^2, \tag{1}$$

and hence by integration we get

$$(1 - \theta)\int_t^s |\xi'|^2 \leq f(\xi(t)) - f(\xi(s)) \tag{2}$$

for any $0 \leq t < s < \infty$. In particular this shows that $f(\xi(t))$ is an increasing function of t; hence since there is a compact K such that $\xi(t) \in K \ \forall t$ we conclude that $\lim_{t \to \infty} f(\xi(t))$ exists, and further, if y is any limit point of $\xi(t)$ as $t \to \infty$, then we must have $\lim_{t \to \infty} f(\xi(t)) = f(y)$ and $f(\xi(t)) \geq f(y) \ \forall t$. Thus (2) gives

$$(1 - \theta)\int_t^\infty |\xi'|^2 \leq f(\xi(t)) - f(y), \tag{3}$$

and by the Lojasiewicz inequality (i) this gives

$$(1 - \theta)\int_t^s |\xi'|^2 \leq |\nabla f(\xi(t))|^{2/(2-\alpha)} \tag{4}$$

provided $|\xi(t) - y| < \eta$. Since $|R| \leq \theta|\xi'|$, we can use the triangle inequality to give $|\nabla f(\xi)| \leq (1 + \theta)|\xi'|$, and hence (4) implies

$$(1 - \theta)\int_t^\infty |\xi'|^2 \leq (1 + \theta)^{2/(2-\alpha)}|\xi'(t)|^{2/(2-\alpha)}, \tag{5}$$

provided $|\xi(t) - y| < \eta$, so that if $|\xi(t) - y| < \eta$ on the interval $[t_1, t_2]$ we can integrate the differential inequality (5) to yield

$$\left(\int_t^\infty |\xi'|^2\right)^{\alpha-1} - \left(\int_{t_1}^\infty |\xi'|^2\right)^{\alpha-1} \geq C(t - t_1), \quad t \in [t_1, t_2],$$

3.10. Uniqueness of Tangent maps with isolated singularities

which evidently implies

(6) $$\int_t^\infty |\xi'|^2 \leq C(t-t_1)^{-1-\beta}, \quad t \in [t_1, t_2], \quad \beta = (1-\alpha)^{-1} - 1 > 0.$$

Notice that by the Cauchy-Schwarz inequality we have

(7) $$|\xi(t_1) - \xi(s)| \leq \left|\int_{t_1}^s \xi'(\tau)\, d\tau\right| \leq (s-t_1)^{1/2} \left(\int_{t_1}^s |\xi'(\tau)|^2 d\tau\right)^{1/2}.$$

Now $\int_t^\infty |\xi'|^2 \to 0$ as $t \to \infty$, and y is a limit point of $\xi(t)$ as $t \to \infty$. We can therefore choose t_1 in the above such that $\int_{t_1}^\infty |\xi'|^2 < \varepsilon^2$ and $|\xi(t_1) - y| < \eta/8$ where ε is a constant to be chosen, depending only on η, α, shortly. Then (7) gives

(8) $$\sup_{t \in [t_1, t_1+T]} |\xi(t_1) - \xi(t)| \leq \varepsilon T^{1/2}.$$

But by multiplying by $(t-t_1)^{\beta/2}$ and integrating in (6) we see that

$$\int_{t_1+T}^{t_1+T_*} (s-t_1)^{1+\beta/2} |\xi'(s)|^2 ds \leq CT^{-\beta/2},$$

with C independent of T for any $T_* > T$ such that $|\xi(t) - \xi(t_1)| < \eta/2$ for all $t \in [t_1+T, t_1+T_*]$. On the other hand then by the Cauchy-Schwarz inequality we get

(9) $|\xi(t_1+T_*) - \xi(t_1+T)| \leq$
$$\leq \int_{t_1+T}^{t_1+T_*} |\xi'(s)|\, ds$$
$$\leq \left(\int_{t_1+T}^{t_1+T_*} (s-t_1)^{1+\beta/2} |\xi'(s)|^2\right)^{1/2} \left(\int_{t_1+T}^{t_1+T_*} (s-t_1)^{-1-\beta/2}\, ds\right)^{1/2}$$
$$\leq CT^{-\beta} \leq \frac{\eta}{4}$$

for $T \geq T_0$, with T_0 fixed, depending only on α and η. Now with this T_0, we choose ε in (8) such that $\varepsilon T_0^{1/2} < \eta/8$, so that (8) gives $|\xi(t_1) - \xi(t_1+T)| \leq \eta/8$ for $T \leq T_0$ and (9) gives $|\xi(t_1+T_*) - \xi(T_1+T)| \leq \eta/4$, hence by the triangle inequality $|\xi(t_1+T_*) - \xi(t_1)| < 3\eta/8$. But then $|\xi(t_1+T_*) - \xi(t_1)| < 3\eta/8$ for arbitrary T_* such that $|\xi(t) - \xi(t_1)| < \eta/2$, $t \in [t_1, T_*]$. Evidently then $|\xi(t_1+t) - \xi(t_1)| < \eta/2$ for all $t \in [t_1, \infty)$. But then (9) can be applied with arbitrary $T > 0$ and with $T_* = t_j$ such that $\xi(t_j) \to y$, thus giving

$$|\xi(t_1+t) - y| \leq Ct^{-\beta/2} \quad \forall t > 0.$$

This completes the proof. □

3.11 Functionals on vector bundles

We begin with a general discussion of the notion of smooth functionals defined on smooth sections of a vector bundle over a compact Riemannian manifold.

Let $n < p_1$, $q \leq p_2$ be positive integers, let Σ be a compact n-dimensional C^1 Riemannian manifold isometrically embedded in $\mathbb{R}^{p_1}$, and let $\mathbf{V} = \cup_{\omega \in \Sigma} V_\omega$, where each V_ω is a q-dimensional subspace of $\mathbb{R}^{p_2}$, with V_ω varying smoothly with respect to ω in the sense that the matrix of the orthogonal projection P_ω of $\mathbb{R}^{p_2}$ onto V_ω is a smooth function of ω.

For the applications to the energy functional, the reader should keep in mind the case when we have $n-1$, p in place of n, p_2 and when $\Sigma = S^{n-1}$, and $V_\omega = T_{\varphi(\omega)} N$, where $\varphi : S^{n-1} \to N$ is a fixed smooth (harmonic) map of S^{n-1} into N.

For each $k = 0, 1, \ldots$, $C^k(\mathbf{V})$ will denote the set of C^k sections of $\mathbf{V}$, meaning that each $u \in C^k(\mathbf{V})$ is a map $u \in C^k(\Sigma; \mathbb{R}^{p_2})$ with $u(\omega) \in V_\omega$ for each $\omega \in \Sigma$. Similarly, for any $\alpha \in (0, 1]$, $C^{k,\alpha}(\mathbf{V})$ denotes the set of $C^{k,\alpha}$ sections of $\mathbf{V}$. Also $L^2(\mathbf{V})$ denotes the subspace of $L^2(\Sigma; \mathbb{R}^{p_2})$ equal to the set of $u = (u^1, \ldots, u^{p_2}) \in L^2(\Sigma; \mathbb{R}^{p_2})$ such that $u(\omega) \in V_\omega$ a.e. $\omega \in \Sigma$. Thus the inner product is the usual L^2 inner product $\langle u, v \rangle_{L^2(\mathbf{V})} = \int_\Sigma u(\omega) \cdot v(\omega) \, d\omega$.

Now let $\tau_1, \ldots, \tau_n$ be a locally defined smoothly varying $\mathbb{R}^{p_1}$-valued functions on Σ which form an orthonormal basis for $T_\omega \Sigma$ at each point $\omega \in \Sigma$ where they are defined. Then for any $u \in C^1(\mathbf{V})$ we define

(i) $$\nabla^{\mathbf{V}} u = \sum_{i=1}^n \tau_i \otimes \nabla^{\mathbf{V}}_{\tau_i} u,$$

where

$$\nabla^{\mathbf{V}}_{\tau_i} u = P_\omega(\nabla_{\tau_i} u),$$

with $\nabla_{\tau_i} u$ denoting the ordinary Euclidean directional derivative of u. Notice that the expression on the right of (i) is actually globally defined on Σ, because it is evidently independent of the particular choice of orthonormal basis $\tau_1, \ldots, \tau_n$, and $\nabla^{\mathbf{V}} u$ takes values in $T_\omega \Sigma \otimes V_\omega \subset \mathbb{R}^{p_1} \otimes \mathbb{R}^{p_2}$. In fact we can get an explicit expression for $\nabla^{\mathbf{V}}$ independent of the particular basis $\tau_1, \ldots, \tau_q$ as follows. Since $u = \sum_{j=1}^{p_2} u^j e_j$ we have $\nabla_{\tau_i} u = \sum_{j=1}^{p_2} \tau_i(u^j) e_j$, and since $\sum_{i=1}^n \tau_i(u^j) \tau_i = \nabla^\Sigma u^j$, where ∇^Σ means gradient operator on Σ, we obtain in place of (i) the alternative identity

(ii) $$\nabla^{\mathbf{V}}_{\tau_i} u = \sum_{j=1}^{p_2} (\nabla^\Sigma u^j) \otimes P_\omega(e_j),$$

which is evidently independent of the particular orthonormal basis $\tau_1, \ldots, \tau_n$. We consider a given smooth real-valued function

$$F = F(\omega, Q), \quad \omega \in \Sigma, \ Q \in \mathbb{R}^{p_1} \times \mathbb{R}^{p_1 p_2}, \quad |Q| < \sigma_0,$$

3.11. Functionals on vector bundles

where $\sigma_0 > 0$ is given. Here we are going to identify $\mathbb{R}^{p_1} \otimes \mathbb{R}^{p_2}$ with $\mathbb{R}^{p_1 p_2}$ in the usual way, via the linear isomorphism induced by the map

$$(x^1, \ldots, x^{p_1}) \otimes (y^1, \ldots, y^{p_2}) \mapsto (x^i y^j)_{i=1,\ldots,p_1, j=1,\ldots,p_2}.$$

(The right side here is a $p_1 \times p_2$ matrix which, since it has $p_1 p_2$ components, can be identified with a point in $\mathbb{R}^{p_1 p_2}$.) Subject to this identification, we can define a functional $\mathcal{F}$ on $C^1(\mathbf{V})$ by

(iii) $$\mathcal{F}(u) = \int_\Omega F(\omega, u, \nabla^\mathbf{V} u), \quad u \in C^1(\mathbf{V}).$$

The Euler-Lagrange operator $\mathcal{M}_\mathcal{F}$ for $\mathcal{F}$ is defined on $C^2(\mathbf{V})$ by the requirements

(iv) $$\mathcal{M}_\mathcal{F}(u) \in C^0(\mathbf{V}), \quad \left.\frac{d}{ds}\mathcal{F}(u+sv)\right|_{s=0} = \langle \mathcal{M}_\mathcal{F}(u), v \rangle_{L^2(\Sigma)}, \quad u, v \in C^2(\mathbf{V}).$$

Of course this does uniquely determine $\mathcal{M}_\mathcal{F}$; indeed to see this and at the same time to get a clear idea of the form of $\mathcal{M}_\mathcal{F}$, we note that by (ii) we can write $F(\omega, u, \nabla^\mathbf{V} u) = \widetilde{F}(\omega, u, \nabla^\Sigma u^1, \ldots, \nabla^\Sigma u^{p_2})$, where $\widetilde{F}(\omega, z, \eta)$, $z \in \mathbb{R}^q$ and $\eta = (\eta^{(1)}, \ldots, \eta^{(p_2)})$, with $\eta^{(\alpha)} = (\eta_1^{(\alpha)}, \ldots, \eta_{p_1}^{(\alpha)}) \in \mathbb{R}^{p_1}$, is defined by

$$\widetilde{F}(\omega, z, \eta^{(1)}, \ldots, \eta^{(p_2)}) = F(\omega, z, \sum_{j=1}^{p_2} \eta^{(j)} \otimes P_\omega(e_j)).$$

Then by direct computation, the derivative on the $\left.\frac{d}{ds}\mathcal{F}(u+sv)\right|_{s=0}$ is given precisely by the expression

$$\left.\frac{d}{ds}\mathcal{F}(u+sv)\right|_{s=0} = \int_\Sigma \sum_{\alpha=1}^{p_2} \left(\sum_{j=1}^{p_1} (\nabla_j v^\alpha) \widetilde{F}_{\eta_j^{(\alpha)}}(\omega, \widetilde{u}, \nabla \widetilde{u}) + v^\alpha \widetilde{F}_{z^\alpha}(\omega, \widetilde{u}, \nabla \widetilde{u}) \right),$$

where $\nabla_j = e_j \cdot \nabla^\Sigma$, $j = 1, \ldots, p_1$ (so that $\nabla^\Sigma f = (\nabla_1 f, \ldots, \nabla_{p_1} f)$) and where the subscripts mean partial derivatives with respect to the indicated variables. After using the integration formula $\int_\Sigma \nabla_j f = - \int_\Sigma f H_j$ on Σ (with $f \in C^\infty(\Sigma)$ and with $(H_1, \ldots, H_{p_1})$ the mean-curvature vector of Σ) this gives

$$\left.\frac{d}{ds}\mathcal{F}(u+sv)\right|_{s=0} =$$
$$= \int_\Sigma \sum_{\alpha=1}^{p_2} v^\alpha \left(-\sum_{j=1}^{p_1} \nabla_j(\widetilde{F}_{\eta_j^{(\alpha)}}(\omega, \widetilde{u}, \nabla \widetilde{u})) + H_j \widetilde{F}_{\eta_j^{(\alpha)}}(\omega, \widetilde{u}, \nabla \widetilde{u}) + \widetilde{F}_{z^\alpha}(\omega, \widetilde{u}, \nabla \widetilde{u}) \right),$$

and hence we see that indeed $\mathcal{M}_\mathcal{F}$ exists and is given by

(v) $$e_\alpha \cdot \mathcal{M}_\mathcal{F}(u) = \sum_{j=1}^{p_1} \nabla_j(F_{\eta_j^{(\alpha)}}(\omega, \widetilde{u}, \nabla \widetilde{u})) - F_{z^\alpha(\omega, \widetilde{u}, \nabla \widetilde{u})} - \sum_{j=1}^{p_1} H_j \widetilde{F}_{\eta_j^{(\alpha)}}(\omega, \widetilde{u}, \nabla \widetilde{u})$$

for each $\alpha = 1, \ldots, p_2$. Notice that this takes the general form

$$e_\alpha \cdot \mathcal{M}_\mathcal{F}(u) = \sum_{i,j=1}^{p_1} \sum_{\beta=1}^{p_2} \widetilde{F}_{\eta_i^{(\alpha)} \eta_j^{(\beta)}}(w, \widetilde{u}, \nabla \widetilde{u}) \nabla_i \nabla_j u^\beta - f_\alpha(w, \widetilde{u}, \nabla \widetilde{u}),$$

where $f = f(w, z, \eta)$ is a smooth function of $(w, z, \eta) \in \Sigma \times \mathbb{R}^{p_2} \times \mathbb{R}^{p_1 p_2}$.

We always assume that F is such that the Euler-Lagrange operator $\mathcal{M}_\mathcal{F}$ is *elliptic*, in the sense that

(vi) $$\sum_{i,j=1}^{p_1} \sum_{\alpha,\beta=1}^{p_2} \widetilde{F}_{\eta_i^{(\alpha)} \eta_j^{(\beta)}}(w, z, \eta) \lambda_\alpha \lambda_\beta \xi^i \xi^j > 0$$

for all $\xi = (\xi^1, \ldots, \xi^{p_1}) \in T_w \Sigma \setminus \{0\}$ and for all $\lambda = (\lambda_1, \ldots, \lambda_{p_2}) \in V_w \setminus \{0\}$.

We also need to mention the linearization of $\mathcal{M}_\mathcal{F}$. If $u \in C^2(\mathbf{V})$ is a solution of $\mathcal{M}_\mathcal{F}(u) = 0$ on Σ, we can define the linearized operator $\mathcal{L}_{\mathcal{F},u}$ of $\mathcal{M}_\mathcal{F}$ at u by

$$\mathcal{L}_{\mathcal{F},u}(v) = \frac{d}{ds} \mathcal{M}_\mathcal{F}(u + sv) \bigg|_{s=0}, \quad v \in C^2(\mathbf{V}).$$

Let $\sigma_0 > 0$ and notice that if $u \in C^2(\mathbf{V})$ with $|u|_{C^2} \leq \sigma_0$ and if u_1, u_2 are arbitrary $C^2(\mathbf{V})$ functions with $|u_1|_{C^2}, |u_2|_{C^2} \leq \sigma_0$, then we can write $\mathcal{M}_\mathcal{F}(u_j) \equiv \mathcal{M}_\mathcal{F}(u + (u_j - u))$ for $j = 1, 2$, and, using the calculus identity $f(1) = f(0) + f'(0) + \int_0^1 (1-s) f''(s) \, ds$ with $f(s) = \mathcal{M}_\mathcal{F}(u + s(u_j - u))$, we deduce

(vii) $$\mathcal{M}_\mathcal{F}(u_j) - \mathcal{M}_\mathcal{F}(u) = \mathcal{L}_{\mathcal{F},u}(u_j - u) + N(u, u_j),$$

where

$$N(u, u_j) = \int_0^1 (1-s) \frac{d^2}{ds^2} \mathcal{M}(u + s(u_j - u)) \, ds, \quad j = 1, 2.$$

By taking the difference of these identities for u_1, u_2 and keeping in mind that $|u + s(u_j - u)|_{C^2} \leq \sigma_0 \ \forall s \in [0, 1]$, we obtain an identity of the form

(viii) $$\mathcal{M}_\mathcal{F}(u_1) - \mathcal{M}_\mathcal{F}(u_2) = \mathcal{L}_{\mathcal{F},u}(u_1 - u_2) + a \cdot \nabla^2(u_1 - u_2) + \\ + b \cdot \nabla(u_1 - u_2) + c \cdot (u_1 - u_2),$$

where

(ix) $$\sup(|a| + |b| + |c|) \leq C(|u_1 - u|_{C^2} + |u_2 - u|_{C^2})$$

with C depending only on $\mathcal{F}$ and σ_0. We shall use (viii), (ix) in the next section.

3.12 The Liapunov-Schmidt Reduction

In this section we want to describe the Liapunov-Schmidt reduction associated with the Euler-Lagrange operator $\mathcal{M}_\mathcal{F}$. The reader unfamiliar with this may find it

3.12. The Liapunov-Schmidt Reduction

instructive to first look at the finite dimensional version in Appendix 3.16 below. The infinite dimensional version described here is exactly analogous, both with respect to the results and the proofs.

Let K be the (finite dimensional) kernel of the elliptic operator $\mathcal{L}_{\mathcal{F}}$ defined in the previous section, let P_K be the orthogonal projection of $L^2(\mathbf{V})$ onto K, and let

(i) $$\mathcal{N}u = P_K u + \mathcal{M}_{\mathcal{F}} u.$$

Notice that then $\mathcal{N}(0) = 0$ (because $\mathcal{M}_{\mathcal{F}}(0) = 0$), and the linearization $d\mathcal{N}|_0(v) \equiv \frac{d}{ds}\mathcal{N}(sv)\big|_{s=0}$ (defined on $C^2(\mathbf{V})$) is just

$$P_K + \mathcal{L}_{\mathcal{F}},$$

which has trivial kernel on $C^2(\mathbf{V})$, so by elliptic theory (keeping in mind the ellipticity 3.11(vi)) we know that $d\mathcal{N}|_0$ is an isomorphism of $C^{2,\alpha}(\mathbf{V})$ onto $C^{0,\alpha}(\mathbf{V})$ for each $\alpha \in (0,1)$. Then the inverse function theorem is applicable to the C^1 operator $\mathcal{N} : C^{2,\alpha}(\mathbf{V}) \to C^{0,\alpha}(\mathbf{V})$, thus giving that $\mathcal{N}$ is a bijection of a neighbourhood U of 0 in $C^{2,\alpha}(\mathbf{V})$ onto a neighbourhood W of 0 in $C^{0,\alpha}(\mathbf{V})$, and that the inverse $\Psi = \mathcal{N}^{-1}$ from W onto U is also C^1. We assume subsequently that $U \subset \{u \in C^{2,\alpha}(\mathbf{V}) : \|u\|_{C^{2,\alpha}} < 1\}$.

Lemma 1 *For a neighbourhood $\widetilde{W} \subset W$ of 0 in $C^{0,\alpha}(\mathbf{V})$, depending only on $\mathcal{F}$, we have*

(ii) $$\|\Psi(f_1) - \Psi(f_2)\|_{W^{2,2}} \leq C\|f_1 - f_2\|_{L^2}, \quad f_1, f_2 \in \widetilde{W},$$

where C depends only on $\mathcal{F}$, and where

$$\|v\|_{W^{2,2}}^2 = \|v\|_{L^2}^2 + \|\nabla v\|_{L^2}^2 + \|\nabla^2 v\|_{L^2}^2.$$

Remark: Of course, since it is merely a notational matter, we can subsequently take $\widetilde{W} = W$.

Proof: Let $u_j = \Psi(f_j)$, so that (since $\mathcal{N}\Psi(f_j) = f_j$) we have

$$P_K(u_j) + \mathcal{M}_{\mathcal{F}}(u_j) = f_j, \quad f_j \in W,$$

and according to (viii) and (ix) of Section 3.11 (with $u = \varphi$) we thus have

(1) $$P_K(u_2 - u_1) + \mathcal{L}_{\mathcal{F},\varphi}(u_1 - u_2) = a \cdot \nabla^2(u_1 - u_2) + \\ + b \cdot \nabla(u_1 - u_2) + c \cdot (u_1 - u_2) + f_2 - f_1,$$

where

(2) $$\sup(|a| + |b| + |c|) \leq C(|u_1 - \varphi|_{C^2} + |u_2 - \varphi|_{C^2}).$$

Taking projections onto K and $K^\perp$ in (1), and keeping in mind that $\mathcal{L}_{\mathcal{F},\varphi}$ takes values in $K^\perp$, we thus have

(3) $$P_K(u_2 - u_1) = P_K(a \cdot \nabla^2(u_1 - u_2) + b \cdot \nabla(u_1 - u_2) + c \cdot (u_1 - u_2) + f_2 - f_1)$$

and

(4)
$$\mathcal{L}_{\mathcal{F},\varphi}\bigl((u_2-u_1)^\perp\bigr) = \bigl(a\cdot\nabla^2(u_1-u_2)+b\cdot\nabla(u_1-u_2)+c\cdot(u_1-u_2)+f_2-f_1\bigr)^\perp.$$

Now according to (vi) of Section 3.11 the operator $\mathcal{L}_{\mathcal{F},\varphi}$ in the equation (4) is elliptic, and hence by the standard L^2 estimates for such equations we have

(5)
$$\|(u_1-u_2)^\perp\|_{W^{2,2}} \leq C\|F\|_{L^2}, \quad C=C(\mathcal{F}),$$

where F is the right side of (4). In view of (2) we thus have

(6)
$$\|(u_1-u_2)^\perp\|_{W^{2,2}} \leq C(|u_1-\varphi|_{C^2}+|u_2-\varphi|_{C^2})\|u_1-u_2\|_{W^{2,2}} + C\|f_1-f_2\|_{L^2}.$$

On the other hand by taking L^2 norms on each side of (3) and using (2) again we evidently have

$$\|P_K(u_1-u_2)\|_{L^2} \leq C(|u_1-\varphi|_{C^2}+|u_2-\varphi|_{C^2})\|u_1-u_2\|_{W^{2,2}} + C\|f_1-f_2\|_{L^2}.$$

Since K is spanned by an orthonormal set $\varphi_1,\ldots,\varphi_q$ of *smooth* functions, this last inequality evidently implies

(7)
$$\|P_K(u_1-u_2)\|_{W^{2,2}} \leq C(|u_1-\varphi|_{C^2}+|u_2-\varphi|_{C^2})\|u_1-u_2\|_{W^{2,2}} + C\|f_1-f_2\|_{L^2}.$$

By adding (6) and (7) we deduce finally

$$\|u_1-u_2\|_{W^{2,2}} \leq C(|u_1-\varphi|_{C^2}+|u_2-\varphi|_{C^2})\|u_1-u_2\|_{W^{2,2}} + C\|f_1-f_2\|_{L^2},$$

and if $|u_1-\varphi|_{C^2}, |u_2-\varphi|_{C^2}$ are then small enough (depending only of $\mathcal{F}$) to ensure that $C(|u_1-\varphi|_{C^2}+|u_2-\varphi|_{C^2}) < \frac{1}{2}$, we thus have (ii) as claimed. $\square$

Now let $\varphi_1,\ldots,\varphi_l$ $(\in C^\infty(\mathbf{V}))$ be an orthonormal basis for K. By definition

(iii)
$$\mathcal{N}\left(\Psi\left(\sum_{j=1}^l \xi^j\varphi_j\right)\right) = \sum_{j=1}^l \xi^j\varphi_j \quad \text{for } \sum_{j=1}^l \xi^j\varphi_j \in W,$$

so in particular

(iv)
$$\left(\mathcal{M}_{\mathcal{F}}\left(\Psi\left(\sum_{j=1}^l \xi^j\varphi_j\right)\right)\right)^\perp = 0 \quad \text{for } \sum_{j=1}^l \xi^j\varphi_j \in W,$$

where $(\,\cdot\,)^\perp$ means $L^2(\mathbf{V})$-orthogonal projection onto the orthogonal complement $K^\perp$ of K. Using (ii) with $f_1 = P_K u$ and $f_2 = \Psi^{-1}(u)$, where u is such that $P_K u \in W$ and $u \in U$, we obtain

(v)
$$\|\Psi(P_K u)-u\|_{W^{2,2}} \leq C\|P_K u - \Psi^{-1}(u)\|_{L^2} \equiv \|\mathcal{M}_{\mathcal{F}} u\|_{L^2}$$

3.12. The Liapunov-Schmidt Reduction

by the definition (i). In particular, with $P_K u$ in place of u this gives

(vi) $\qquad \|\Psi(P_K u) - P_K u\|_{W^{2,2}} \leq C \|P_K u\|_{L^2}^2, \quad u \in \widetilde{U},$

where $\widetilde{U} = \{u \in U : P_K u \in W\} = U \cap P_K^{-1}(W \cap K)$, because $\mathcal{L}_{\mathcal{F}} P_K u = 0$ and hence $\|\mathcal{M}_{\mathcal{F}}(P_K u)\|_{L^\infty} \leq C \|P_K u\|_{C^2}^2 \leq C \|P_K u\|_{L^2}^2$. (Notice that trivially $\|P_K u\|_{C^2} \leq C \|P_K u\|_{L^2}$, because the φ_j, $j = 1, \ldots, l$, are all $C^\infty(\mathbf{V})$ functions.) In particular (vi) evidently implies

(vii) $\qquad d\Psi|_0 \circ P_K = P_K.$

We now define

$$f(\xi) = \mathcal{F}\bigl(\Psi(\textstyle\sum_{j=1}^{l} \xi^j \varphi_j)\bigr), \quad \textstyle\sum_{j=1}^{l} \xi^j \varphi_j \in W,$$

and check by direct computation (using the definition 3.11(iv) of $\mathcal{M}_{\mathcal{F}}$) that

(viii) $\qquad \langle \eta, \nabla f(\xi) \rangle_{L^2} \equiv \langle \mathcal{M}_{\mathcal{F}}\bigl(\Psi(\textstyle\sum_{j=1}^{l} \xi^j \varphi_j)\bigr), d\Psi(\sum_j \xi^j \varphi_j)(\sum_j \eta^j \varphi_j) \rangle_{L^2}$

for $\xi \in \widetilde{W}$ and $\eta \in \mathbb{R}^q$, where $\widetilde{W}$ is the open neighbourhood of 0 in $\mathbb{R}^l$ such that $\xi \in \widetilde{W} \iff \sum_{j=1}^{l} \xi^j \varphi_j \in W \cap K$. Notice that this can be written

$$\langle \eta, \nabla f(\xi) \rangle_{L^2} \equiv \langle \mathcal{M}_{\mathcal{F}}\bigl(\Psi(\textstyle\sum_{j=1}^{l} \xi^j \varphi_j)\bigr), d\Psi(\sum_j \xi^j \varphi_j)(\sum_j \eta^j \varphi_j) - P_K(\textstyle\sum_{j=1}^{l} \eta^j \varphi_j) \rangle_{L^2} + \langle \mathcal{M}_{\mathcal{F}}\bigl(\Psi(\textstyle\sum_{j=1}^{l} \xi^j \varphi_j)\bigr), P_K(\textstyle\sum_{j=1}^{l} \eta^j \varphi_j) \rangle_{L^2}.$$

Now by (vii) we know that $\|d\Psi(\sum_j \xi^j \varphi_j)(\sum_j \eta^j \varphi_j) - P_K(\sum_{j=1}^{l} \eta^j \varphi_j)\|_{L^2} \leq C|\xi|$; then, taking (by (iv)) $\sum_j \eta^j \varphi_j$ parallel to $\mathcal{M}_{\mathcal{F}}(\Psi(\sum_j \xi^j \varphi_j))$, we deduce that

$$\|\mathcal{M}_{\mathcal{F}}\bigl(\Psi(\textstyle\sum_{j=1}^{l} \xi^j \varphi_j)\bigr)\| \leq |\nabla f(\xi)| + C|\xi| \, \|\mathcal{M}_{\mathcal{F}}\bigl(\Psi(\textstyle\sum_{j=1}^{l} \xi^j \varphi_j)\bigr)\|.$$

Similarly, taking η parallel to $\nabla f(\xi)$ we get

$$|\nabla f(\xi)| \leq (1 + C|\xi|) \, \|\mathcal{M}_{\mathcal{F}}\bigl(\Psi(\textstyle\sum_{j=1}^{l} \xi^j \varphi_j)\bigr)\|.$$

Evidently then, since we may take a smaller neighbourhood W if necessary (to ensure that $C|\xi| \leq 1$), we conclude

(ix) $\qquad \dfrac{1}{2} |\nabla f(\xi)| \leq \|\mathcal{M}_{\mathcal{F}}\bigl(\Psi(\textstyle\sum_{j=1}^{l} \xi^j \varphi_j)\bigr)\| \leq 2|\nabla f(\xi)|, \quad \xi \in \widetilde{W}.$

Also, if $u \in \widetilde{U} \equiv U \cap P_K^{-1}(W \cap K)$ we have

(x) $\qquad \mathcal{M}_{\mathcal{F}} u = 0 \Rightarrow \mathcal{N}(u) = P_K u \Rightarrow \Psi \mathcal{N}(u) = \Psi(P_K u) \Rightarrow u = \Psi(P_K u),$

and also then by the left side of (ix) $\nabla f(\xi) = 0$ with $\xi \in \widetilde{W}$ such that $\sum \xi_j \varphi_j = P_K u$. Conversely, if $\xi \in \widetilde{W}$ and $\nabla f(\xi) = 0$, then $\mathcal{M}(\Psi(\sum \xi_j \varphi_j)) = 0$ by (ix), and hence $P_K(\Psi(\sum \xi_j \varphi_j)) = \mathcal{N}(\Psi(\sum \xi_j \varphi_j)) = \sum \xi_j \varphi_j$ so that automatically $\Psi(\sum \xi_j \varphi_j) \in \widetilde{U}$ if $\xi \in \widetilde{W}$. Thus

(xi) $\qquad \{u \in \widetilde{U} : \mathcal{M}_{\mathcal{F}} u = 0\} = \Psi\bigl(\{\textstyle\sum_{j=1}^{l} \xi^j \varphi_j : \xi \in \widetilde{W} \text{ and } \nabla f(\xi) = 0\}\bigr).$

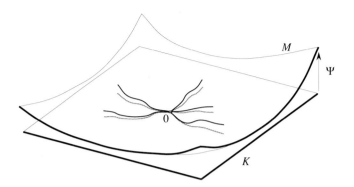

Figure 3.6: *The Liapunov-Schmidt Reduction*

Since Ψ is a C^1 diffeomorphism, we see that Ψ embeds $W \cap K$ into U, so

(xii) $$M := \left\{ \Psi\left(\sum_{j=1}^{l} \xi^j \varphi_j \right) : \xi \in \widetilde{W} \right\}$$

is an l-dimensional embedded C^1 submanifold of U which according to (xi) contains the whole zero set of $\mathcal{M}_{\mathcal{F}}$ in the neighbourhood $\widetilde{U}$. (See Figure 3.6.)

We also note some important facts about how well the function f approximates the functional $\mathcal{F}$ near zero. Specifically, note that for $u \in U$ and $P_K u \in W$ we have

$$\begin{aligned} |\mathcal{F}(u) - \mathcal{F}(\Psi(P_K u))| &= \left| \int_0^1 \frac{d}{ds} \mathcal{F}(u + s(\Psi(P_K u) - u)) ds \right| \\ &= \langle \mathcal{M}_{\mathcal{F}}(u + s(\Psi(P_K u) - u)), \Psi(P_K u) - u \rangle_{L^2(\mathbf{V})} \end{aligned}$$

by the definition 3.11(iv) of $\mathcal{M}_{\mathcal{F}}$. Since

$$\|\mathcal{M}_{\mathcal{F}}(u + s(\Psi(P_K u) - u)) - \mathcal{M}_{\mathcal{F}}(u)\|_{L^2} \leq C \|\Psi(P_K u) - u\|_{W^{2,2}}$$

(by direct computation using the expression 3.11(v)) we thus get

$$|\mathcal{F}(u) - \mathcal{F}(\Psi(P_K u))| \leq \|\mathcal{M}_{\mathcal{F}}(u)\|_{L^2} \|\Psi(P_K u) - u\|_{L^2} + C\|\Psi(P_K u) - u\|_{W^{2,2}}^2 ,$$

and hence by (v) we have

(xiii) $$|\mathcal{F}(u) - \mathcal{F}(\Psi(P_K u))| \leq C \|\mathcal{M}_{\mathcal{F}} u\|_{L^2}^2 .$$

3.13 The Łojasiewicz Inequality for $\mathcal{F}$

We now specialize to the case when the functional $\mathcal{F}$ is real-analytic. To be precise we assume that $F = F(\omega, z, \eta)$, $\omega \in \Sigma$, $z \in \mathbb{R}^{p_2}$, $\eta \in \mathbb{R}^{p_1 p_2}$ is smooth and that all derivatives with respect to the ω variables up to order 3 are real-analytic functions

3.13. The Łojasiewicz Inequality for $\mathcal{F}$

of the variables z, η: Thus we assume that for each $(z_0, \eta_0) \in \mathbb{R}^{p_2} \times \mathbb{R}^{p_1 p_2}$ and each integer $j = 0, 1, 2, 3$ there are smooth functions $\{a_{\alpha\beta}\}$ on Σ corresponding to arbitrary multi-indices $\alpha = (\alpha_1, \ldots, \alpha_{p_1})$, $\beta = (\beta_1, \ldots, \beta_{p_1 p_2})$, and $\sigma > 0$, such that

$$\sum_{m=0}^{\infty} \Big(\sum_{|\alpha|+|\beta|=m} \sup_{\omega \in \Sigma} |D^j_\omega a_{\alpha\beta}(\omega)| \Big) \sigma^m < \infty$$

and such that

$$F(\omega, z, \eta) = \sum_{m=0}^{\infty} \sum_{|\alpha|+|\beta|=m} a_{\alpha\beta}(\omega)(z-z_0)^\alpha (\eta-\eta_0)^\beta$$

for $|z-z_0|+|\eta-\eta_0| < \sigma$. Notice that then we can apply the implicit function theorem argument above on the complexified spaces $\mathbb{C} \otimes C^{2,\alpha}(\mathbf{V})$ and $\mathbb{C} \otimes C^{0,\alpha}(\mathbf{V})$ thus giving $\Psi = \mathcal{N}^{-1}$ defined and C^1 on a neighbourhood $U_\mathbb{C}$ of 0 in the space $\mathbb{C} \otimes C^{0,\alpha}(\mathbf{V})$ in the sense that for any fixed $u_j \in C^{0,\alpha}(\mathbf{V})$, $j = 1, \ldots, R$, the complex derivatives $\frac{\partial}{\partial z^k} \Psi(\sum_{j=1}^R z^j u_j)$ all exist and are continuous as maps from $U_\mathbb{C}$ into $C^{2,\alpha}_\mathbb{C}(\mathbf{V})$. In particular

$$f_\mathbb{C}(z^1, \ldots, z^n) := \mathcal{F}(\Psi(\sum_{j=1}^l z^j \varphi_j))$$

is a holomorphic function of $(z^1, \ldots, z^n)$ in some neighbourhood of 0 in $\mathbb{C}^l$. Thus the original $f(\xi)$ $(= \mathcal{F}(\Psi(\sum_{j=1}^l \xi^j \varphi_j)))$ is real-analytic in some neighbourhood of 0 in $\mathbb{R}^l$. Thus we can apply the Lojasiewicz result 3.10(i) in order to deduce that there exist constants $\alpha \in (0,1]$ and $C, \sigma > 0$ such that

(i) $\qquad |f(\xi)|^{1-\alpha/2} \leq C|\nabla f(\xi)| \quad \forall \xi \in B_\sigma(0)\,,$

where $B_\sigma(0)$ is the ball in $\mathbb{R}^l$ of radius σ and centre 0. But now by the inequality 3.12(xiii), with $u \in U$ arbitrary such that $\xi^j = \langle u, \varphi_j \rangle_{L^2}$ satisfy $|\xi| < \sigma$, we have

$$|\mathcal{F}(u) - f(\xi)| \leq C \|\mathcal{M}_\mathcal{F}(u)\|^2_{L^2}$$

and hence by (i) and 3.12(ix) we then have

$$\begin{aligned} |\mathcal{F}(u)|^{1-\alpha/2} &\leq C(\|\mathcal{M}_\mathcal{F}(u)\|_{L^2} + \|\mathcal{M}_\mathcal{F}(u)\|^{2-\alpha}) \\ &\leq C \|\mathcal{M}_\mathcal{F}(u)\|_{L^2} \end{aligned}$$

for each $u \in U$ such that $\|P_K u\| < \sigma$. In particular there is $\sigma_0 > 0$ such that

(ii) $\qquad |\mathcal{F}(u)|^{1-\alpha/2} \leq C \|\mathcal{M}_\mathcal{F} u\|_{L^2} \quad \forall u \in C^3(\mathbf{V}),\ \|u\|_{C^3} < \sigma_0\,,$

and this is the required Lojasiewicz inequality in $C^3(\mathbf{V})$.

Now we want to prove that, even *without* any real-analyticity hypothesis as in (i), the above inequality holds with best exponent $\alpha = 1$ in case we assume the following "integrability condition":

(iii) $\qquad \mathcal{F}(\Psi(P_K u)) \equiv 0$ on some C^3-neighbourhood of 0 in $C^3(\mathbf{V})$.

This is called an integrability condition because (by 3.12(xi)) it is equivalent to the hypothesis that there is a smooth ℓ-dimensional manifold of solutions of the non-linear equation $\mathcal{M}_\mathcal{F}(u) = 0$ which is tangent to the kernel K of $\mathcal{L}_\mathcal{F}$ at 0.

Thus we claim:

Lemma 1 *If (iii) holds then there are σ, $C > 0$ such that*

$$|\mathcal{F}(u)|^{1/2} \leq C\|\mathcal{M}_\mathcal{F}(u)\|_{L^2}, \quad u \in C^3(\mathbf{V}), \ \|u\|_{C^3} < \sigma.$$

Proof: According to (iii) and 3.12(xiii) we have

$$|\mathcal{F}(u)| = |\mathcal{F}(u) - \mathcal{F}(\Psi(P_K u))| \leq C\|\mathcal{M}_\mathcal{F}(u)\|_{L^2}^2$$

for each $u \in U$, and this is the required inequality. $\square$

3.14 Łojasiewicz for the Energy functional on S^{n-1}

We would like to apply the Łojasiewicz inequality to the energy functional on the sphere, but the smooth maps $C^\infty(S^{n-1}; N)$ are not a linear space, so we first need to show that, at least for maps which are C^3 close to a given harmonic map $\varphi_0 \in C^\infty(S^{n-1}; N)$ we can write the energy functional as a functional on a vector bundle as in Section 3.11.

So take fixed $\varphi_0 \in C^\infty(S^{n-1}; N)$ which is harmonic; recall that by the discussion of Section 2.2 (and also Section 3.10) this is equivalent to φ_0 satisfying the equation

(i) $$\Delta_{S^{n-1}}\varphi_0 + A_{\varphi_0}(D_\omega \varphi_0, D_\omega \varphi_0) = 0,$$

where $A_{\varphi_0}(D_\omega \varphi_0, D_\omega \varphi_0)$ is an abbreviation for $\sum_{j=1}^{n-1} A_{\varphi_0}(\nabla_{\tau_j}\varphi_0, \nabla_{\tau_j}\varphi_0)$ with an orthonormal basis $\tau_1, \ldots, \tau_{n-1}$ for $T_\omega S^{n-1}$. Now for $\delta > 0$ and $y_0 \in N$ we let

$$\begin{aligned} T_\delta(y_0) &= \{\tau \in T_{y_0} N : |\tau| < \delta\} \\ U_\delta(y_0) &= \{\Pi(y_0 + \tau) : \tau \in T_\delta(y_0)\}. \end{aligned}$$

Then, for suitable $\delta = \delta(N) > 0$, $U_\delta(y_0)$ is a neighbourhood of y_0 in N, and in fact the mapping $\Phi_{y_0} : \tau \mapsto \Pi(y_0 + \tau)$ is a smooth (real analytic if N is real analytic) diffeomorphism of $T_\delta(y_0)$ onto $U_\delta(y_0)$ such that

(ii) $\Phi_{y_0}^{-1}(y)$ depends smoothly or real analytically on $(y_0, y) \in N \times N$

according as N is smooth or real-analytic respectively. Now let $\mathbf{V} = \varphi_0^* TN \equiv \{T_{\varphi_0(\omega)} N\}_{\omega \in S^{n-1}}$. If $u \in C^2(S^{n-1}; N)$ with $\|u - \varphi_0\|_{C^2} < \delta$, then by definition of Φ_{y_0} we have the identity

(iii) $$u(\omega) = \Pi(\varphi_0 + \Phi_{\varphi_0(\omega)}^{-1}(u(\omega))), \quad \omega \in S^{n-1},$$

3.14. Łojasiewicz for the Energy functional on S^{n-1}

so, letting $\widetilde{u}(\omega) = \Phi^{-1}_{\varphi_0(\omega)}(u(\omega))$ we have $\widetilde{u} \in C^2(\mathbf{V})$ with

(iv) $$u = \Pi(\varphi_0 + \widetilde{u}).$$

Thus for $u \in C^2(S^{n-1}; N)$ we have the identity

(v) $$\mathcal{E}_{S^{n-1}}(u) = \mathcal{E}_{S^{n-1}}(\Pi(\varphi_0 + \widetilde{u})).$$

Notice that the expression on the right has the form
$$\int_{S^{n-1}} F(\omega, \widetilde{u}, \nabla^{\mathbf{V}} \widetilde{u}),$$

where we use the notation of Section 3.11 with $\mathbf{V} = \{V_\omega\}_{\omega \in S^{n-1}}$, $V_\omega = T_{\varphi_0(\omega)}N$ and where
$$F = F(\omega, z, \eta), \quad \omega \in S^{n-1}, \, z \in \mathbb{R}^p, \, \eta \in \mathbb{R}^{np}$$

is given (see 3.11(ii)) by

(vi) $$F(\omega, z, \eta) = \left| d\Pi|_{\varphi_0(\omega)+z} \left(\left(\sum_{\alpha=1}^{p} \nabla^{S^{n-1}} \varphi_0^\alpha(\omega) \otimes P_\omega e_\alpha \right) + \eta \right) \right|^2,$$

which is certainly a C^∞ function of ω, z, η for $|z| < \delta$ and which also satisfies the analyticity condition of Section 3.13 in case N is real-analytic. (Because Π is C^∞ or real-analytic in $\{x \in \mathbb{R}^p : \operatorname{dist}(x, N) < \delta\}$ according as N is C^∞ or real-analytic.) Thus we are motivated to make the following definition: Let $\mathcal{F}$ be defined on $\{u \in C^2(\mathbf{V}) : \|u\|_{C^2} < \delta\}$ by

(vii) $$\mathcal{F}(u) = \int_{S^{n-1}} (F(\omega, u, \nabla^{\mathbf{V}} u) - F(\omega, 0, 0)),$$

with F as in (vi). Then in view of (v), $\mathcal{F}$ is related to the energy functional by

(viii) $$\mathcal{E}_{S^{n-1}}(u) - \mathcal{E}_{S^{n-1}}(\varphi_0) = \mathcal{F}(\widetilde{u}) = \mathcal{E}_{S^{n-1}}(\Pi(\varphi_0 + \widetilde{u})) - \mathcal{E}_{S^{n-1}}(\varphi_0)$$

for $u \in C^2(S^{n-1}; N)$ and $\widetilde{u}$ such that $u = \Pi(\varphi_0 + \widetilde{u})$ as described above; stated in another way, this says

(ix) $$\mathcal{F}(v) = \mathcal{E}_{S^{n-1}}(\Pi(\varphi_0 + v)) - \mathcal{E}_{S^{n-1}}(\varphi_0), \qquad v \in C^2(\mathbf{V}), \, \|v\|_{C^2} \leq \delta.$$

Next, recall (see the discussion of Section 2.2) that the Euler-Lagrange operator of $\mathcal{E}_{S^{n-1}}(u)$ is exactly $\mathcal{M}_{\mathcal{E}_{S^{n-1}}}(u) = \Delta_{S^{n-1}} u + A_u(D_\omega u, D_\omega u)$, and it satisfies the identity

$$\frac{d}{ds} \mathcal{E}_{S^{n-1}}(\Pi(u + sv))|_{s=0} = -\langle \mathcal{M}_{\mathcal{E}_{S^{n-1}}}(u), v \rangle_{L^2}, \quad v = (v^1, \dots, v^p) \in C^2(S^{n-1}; \mathbb{R}^p).$$

On the other hand by (viii) and the definition $\frac{d}{ds}\mathcal{F}(u+sv)|_{s=0} = -\langle \mathcal{M}_{\mathcal{F}}(u), v \rangle_{L^2}$ we then have

$$\langle \mathcal{M}_{\mathcal{E}_{S^{n-1}}}(u), v \rangle_{L^2} = \langle \mathcal{M}_{\mathcal{F}}(\widetilde{u}), v \rangle_{L^2}, \qquad v \in C^2(\mathbf{V}),$$

which evidently implies

(x) $\quad (\mathcal{M}_{\mathcal{E}_{S^{n-1}}}(u))^{T_0} \equiv \mathcal{M}_{\mathcal{F}}(\tilde{u}), \quad u \in C^2(S^{n-1}; N), \ \|u - \varphi_0\|_{C^2} < \delta,$

where $(\cdot)^{T_0}$ means the orthogonal projection onto $T_{\varphi_0(\omega)} N$. However recall that $\mathcal{M}_{\mathcal{E}_{S^{n-1}}}(u) = (\Delta u)^T$ (see (iv') of Section 2.2), where $(\cdot)^T$ means orthogonal projection onto $T_{u(\omega)} N$. Since $\|u - \varphi_0\|_{C^2} < \delta$ we then have from (ix) the pointwise bounds

(xi) $\quad (1 - C\delta)|\mathcal{M}_{\mathcal{E}_{S^{n-1}}}(u)| \leq |\mathcal{M}_{\mathcal{F}}(\tilde{u})| \leq |\mathcal{M}_{\mathcal{E}_{S^{n-1}}}(u)|.$

Notice that by taking the mixed partial derivative $\frac{\partial^2}{\partial s \partial t}$ of the identity (ix) with $sv + tw$ in place of v, we also have

$$\mathcal{L}_{\varphi_0}(v) \equiv \mathcal{L}_{\mathcal{F}}(v),$$

where $\mathcal{L}_{\varphi_0} = \frac{d}{ds} \mathcal{M}_{\mathcal{E}_{S^{n-1}}}(\Pi(\varphi_0 + sv))|_{s=0}$ is the linearization of $\mathcal{M}_{\mathcal{E}_{S^{n-1}}}$ at φ_0 and $\mathcal{L}_{\mathcal{F}}(v) = \frac{d}{ds} \mathcal{M}_{\mathcal{F}}(sv)|_{s=0}$ is the linearization of $\mathcal{M}_{\mathcal{F}}$ at 0; in particular $\mathcal{L}_{\mathcal{F}}$ is certainly elliptic—it has second-order term Δv—and hence we may apply all the theory of the Sections 3.11—3.13 to $\mathcal{F}$.

Thus in particular if N is real-analytic (which ensures $\mathcal{F}$ is real-analytic by the discussion above), we know by (xi) and the Lojasiewicz inequality 3.13(ii) that there is $\alpha \in (0, 1]$ and $C, \sigma > 0$ such that

(xii) $\quad |\mathcal{E}_{S^{n-1}}(u) - \mathcal{E}_{S^{n-1}}(\varphi_0)|^{1-\alpha/2} \leq C \|\mathcal{M}_{\mathcal{E}_{S^{n-1}}}(u)\|_{L^2},$
for $u \in C^\infty(S^{n-1}; N)$ with $\|u - \varphi_0\|_{C^3} < \sigma$.

If on the other hand we have that N is merely smooth, then assuming that $\mathcal{F}$ satisfies the integrability condition 3.13(iii)

(xiii) $\quad \mathcal{F}(\Psi(P_K u)) \equiv 0$ on some C^3-neighbourhood of 0 in $C^3(\mathbf{V})$.

(which by (ix) and the discussion in Section 3.13 is the same as the requirement that the set of $u \in C^\infty(S^{n-1}; N)$ with $\|u - \varphi_0\|_{C^3} < \delta$ and $\mathcal{M}_{\mathcal{E}_{S^{n-1}}}(u) = 0$ is an ℓ-dimensional submanifold), we have (xii) with best exponent $\alpha = 1$; that is, there are $C, \sigma > 0$ such that

(xiv)
$|\mathcal{E}_{S^{n-1}}(u) - \mathcal{E}_{S^{n-1}}(\varphi_0)|^{1/2} \leq C \|\mathcal{M}_{\mathcal{E}_{S^{n-1}}}(u)\|_{L^2}, \ u \in C^\infty(S^{n-1}; N), \ \|u - \varphi_0\|_{C^3} < \sigma.$

3.15 Proof of Theorem 1 of Section 3.10

The proof here is a simplification of the original proof in [Si83b].

First recall, by definition of tangent map, that there is a sequence $\rho_j \downarrow 0$ such that the re-scaled mappings $u_{y,\rho_j} (= u(y + \rho_j x))$ converge in the $W^{1,2}$-norm to our tangent

3.15. Proof of Theorem 1 of Section 3.10

map φ. Thus for any given $\eta > 0$, for suitable ρ, for example for $\rho = \rho_j$ with j sufficiently large depending on η, we have

(1) $$\int_{B_{3/2} \setminus B_{3/4}} |u_{y,\rho} - \varphi|^2 < \eta^2,$$

where, here and subsequently, B_ρ is an abbreviation for $B_\rho(0)$. Let us now abbreviate $\tilde{u} = u_{y,\rho}$, and keep this ρ fixed for the time being, and also small enough so that $\overline{B}_{3\rho/2} \subset \Omega$; thus $\tilde{u}$ is at least defined on $\overline{B}_{3/2}$. Now since φ is homogeneous in $\mathbb{R}^n \setminus \{0\}$ and smooth on S^{n-1} it is clear that if $B_\sigma(z) \subset B_{3/2} \setminus B_{3/4}$ then

$$\sigma^{-n} \int_{B_\sigma(z)} |\tilde{u} - \varphi(z)|^2 \leq 2\sigma^{-n} \int_{B_\sigma(z)} |\tilde{u} - \varphi|^2 + 2\sigma^{-n} \int_{B_\sigma(z)} |\varphi - \varphi(z)|^2$$
$$\leq 2\sigma^{-n}\eta^2 + \beta\sigma^2,$$

where β is a fixed constant depending on φ but not depending on σ or ρ. Thus if $\gamma > 0$ is given, then for small enough η, σ (depending only on n, N, φ, γ) we can apply the ε-regularity theorem on the ball $B_\sigma(z)$ in order to deduce that $\|u - \varphi\|_{C^3(B_{\sigma/2}(z))} \leq \gamma$. Thus (in view of the arbitrariness of z) we obtain for any given $\gamma > 0$ that there exists $\eta = \eta(\gamma, \varphi) > 0$ such that

(2) $$\|\tilde{u} - \varphi\|_{L^2(B_{3/2} \setminus B_{3/4})} < \eta \Longrightarrow \|\tilde{u} - \varphi\|_{C^3(B_{5/4} \setminus B_{7/8})} < \gamma.$$

(Notice that in (2) we do *not* have to assume that ρ is proportionately close to one of the ρ_j.)

Next recall that by (xii) we have constants $C > 0$, $\gamma \in (0,1)$ and $\alpha \in (0,1]$ such that

(3) $$\left| \int_{S^{n-1}} (|D_\omega w|^2 - |D_\omega \varphi|^2) \right|^{1-\alpha/2} \leq C \|\mathcal{M}_{\mathcal{E}_{S^{n-1}}}(w)\|_{L^2}, \qquad \|w - \varphi\|_{C^3} < \gamma.$$

Since $\|\mathcal{M}_{\mathcal{E}_{S^{n-1}}}(w)\|_{L^2}$ is uniformly bounded for $\|w\|_{C^2} \leq 1$ (so that trivially we have $\|\mathcal{M}_{\mathcal{E}_{S^{n-1}}}(w)\|_{L^2}^{\gamma_2} \leq C^{\gamma_2 - \gamma_1} \|\mathcal{M}_{\mathcal{E}_{S^{n-1}}}(w)\|_{L^2}^{\gamma_1}$ for $\gamma_1 \leq \gamma_2$), we can, and we shall, assume $\alpha < 1$.

So from now on $\alpha \in (0,1), \gamma$ (depending on φ) are chosen fixed such that (3) holds, and η, depending on φ and γ, is chosen so that the implication (2) holds.

Now by the monotonicity identity 2.4(iii) we have

(4) $$2 \int_{B_1} r^{2-n} \left| \frac{\partial \tilde{u}}{\partial r} \right|^2 = \int_{B_1} |D\tilde{u}|^2 - \Theta_{\tilde{u}}(0).$$

Notice that since $\tilde{u} = u_{y,\rho}$ and $\int_{B_1} r^{2-n} \left| \frac{\partial \tilde{u}}{\partial r} \right|^2 = \int_{B_\rho(y)} r^{2-n} \left| \frac{\partial u}{\partial r} \right|^2 \to 0$ as $\rho \downarrow 0$, we have

(5) $$\int_{B_1} r^{2-n} \left| \frac{\partial \tilde{u}}{\partial r} \right|^2 \leq 1$$

for all $\rho \leq \rho_0$ for suitable $\rho_0 > 0$. Also, in proving the identity (4) (in Section 2.4) we showed that $\tilde{u}$ satisfied the identity

$$(n-2)\int_{B_1}|D\tilde{u}|^2 = \int_{\partial B_1}\left(|D\tilde{u}|^2 - 2\left|\frac{\partial \tilde{u}}{\partial r}\right|^2\right) \leq \int_{\partial B_1}|D'\tilde{u}|^2,$$

where D' means tangential gradient on ∂B_1. Then using this in (4), and keeping in mind that

$$\Theta_{\tilde{u}}(0) = \Theta_u(y) = \Theta_\varphi(0) = \int_{B_1}|D\varphi|^2 = \frac{1}{n-2}\int_{S^{n-1}}|D_\omega\varphi|^2$$

by virtue of the fact that φ is a tangent map of u at y (hence homogeneous of degree zero), we obtain

$$2(n-2)\int_{B_1}r^{2-n}\left|\frac{\partial \tilde{u}}{\partial r}\right|^2 \leq \int_{\partial B_1}(|D'\tilde{u}|^2 - |D'\varphi|^2)$$
$$= \int_{S^{n-1}}(|D_\omega\tilde{u}|^2 - |D_\omega\varphi|^2).$$

Now in view of (2) we can apply the inequality (3) in order to deduce

(6) $$\int_{B_1}r^{2-n}\left|\frac{\partial \tilde{u}}{\partial r}\right|^2 \leq C\|\mathcal{M}_{\mathcal{E}_{S^{n-1}}}(\tilde{u})\|_{L^2}^{1/(1-\alpha/2)},$$

so long as $\|\tilde{u} - \varphi\|_{L^2(B_{3/2}\setminus B_{3/4})} < \eta$.

Now $\tilde{u}$ satisfies the equation $\Delta \tilde{u} + \sum_{j=1}^n A_{\tilde{u}}(D_j\tilde{u}, D_j\tilde{u}) = 0$, which, in terms of spherical coordinates $r = |x|$, $\omega = |x|^{-1}x$, can be written

$$\frac{1}{r^{n-1}}\frac{\partial}{\partial r}\left(r^{n-1}\frac{\partial \tilde{u}}{\partial r}\right) + \frac{1}{r^2}\Delta_{S^{n-1}}\tilde{u} + \frac{1}{r^2}A_{\tilde{u}}(D_\omega\tilde{u}, D_\omega\tilde{u}) + A_{\tilde{u}}\left(\frac{\partial \tilde{u}}{\partial r}, \frac{\partial \tilde{u}}{\partial r}\right) = 0.$$

Since this just says $\mathcal{M}_{\mathcal{E}_{S^{n-1}}}(\tilde{u}) = -\frac{1}{r^{n-1}}\frac{\partial}{\partial r}(r^{n-1}\frac{\partial \tilde{u}}{\partial r}) - A_{\tilde{u}}(\frac{\partial \tilde{u}}{\partial r}, \frac{\partial \tilde{u}}{\partial r})$, we see that (6) implies

(7) $$\left(\int_{B_1}r^{2-n}\left|\frac{\partial \tilde{u}}{\partial r}\right|^2\right)^{2-\alpha} \leq C\int_{S^{n-1}}\left(\left|\frac{\partial(r^{n-1}\tilde{u})}{\partial r}\right|^2 + \left|\frac{\partial \tilde{u}}{\partial r}\right|^2\right),$$

provided that $\|\tilde{u} - \varphi\|_{L^2(B_{3/2}\setminus B_{3/4})} < \eta$.

Now notice that the re-scaled function $\tilde{u}^{(\sigma)}$ defined by $\tilde{u}^{(\sigma)}(x) = \tilde{u}(\sigma x)$ also satisfies the harmonic map equation

$$\Delta \tilde{u}^{(\sigma)} + \sum_{j=1}^n A_{\tilde{u}^{(\sigma)}}(D_j\tilde{u}^{(\sigma)}, D_j\tilde{u}^{(\sigma)}) = 0,$$

and by differentiation with respect to σ, and noting that $\frac{\partial \tilde{u}(\sigma x)}{\partial \sigma} = x \cdot D\tilde{u}(\sigma x) = \frac{r}{\sigma}\frac{\partial \tilde{u}(\sigma)}{\partial r}$, we obtain the linear equation

$$\mathcal{L}\left(r\frac{\partial \tilde{u}}{\partial r}\right) = 0,$$

3.15. Proof of Theorem 1 of Section 3.10

where $\mathcal{L}$ is the linear elliptic operator obtained by linearizing $\mathcal{M}(w) \equiv \Delta w + \sum_{j=1}^{n} A_w(D_j w, D_j w)$ at $w = \tilde{u}$. Thus

$$\mathcal{L}v = \Delta v + 2\sum_{j=1}^{n} A_{\tilde{u}}(D_j v, D_j \tilde{u}) + \sum_{j=1}^{n}\sum_{k=1}^{p} v^k \left.\frac{\partial A_z}{\partial z^k}\right|_{z=\tilde{u}} (D_j \tilde{u}, D_j \tilde{u}) = 0.$$

Now since $\|\tilde{u} - \varphi\|_{C^3(B_{5/4}\setminus B_{7/8})} \leq \gamma \leq 1$, we see that this operator has the form

$$\mathcal{L}(v) = \Delta v + b \cdot Dv + c \cdot v,$$

where $|b| + |c| \leq \beta$, $\beta = \beta(n, N)$, in the domain $\Omega = B_{5/4} \setminus B_{7/8}$. But any solution $v = (v^1, \ldots, v^p)$ of $\mathcal{L}(v) = 0$ for such an operator $\mathcal{L}$ satisfies the estimate

$$\sup_{B_{\tau/2}(z)} |Dv| \leq C\|v\|_{L^2(B_\tau(z))}$$

for any ball $B_\tau(z) \subset \Omega \ (= B_{5/4} \setminus B_{7/8})$. (See e.g. [GT83, Chapter 8].) Thus, in particular, covering S^{n-1} by a family of such balls $B_{\tau/2}$ with $\tau = 1/16$, we conclude

$$\sup_{S^{n-1}} \left|D\left(r\frac{\partial \tilde{u}}{\partial r}\right)\right|^2 \leq C \int_{B_{5/4}\setminus B_{7/8}} \left|\frac{\partial \tilde{u}}{\partial r}\right|^2,$$

provided that $\|\tilde{u} - \varphi\|_{L^2(B_{3/2}\setminus B_{3/4})} < \eta$. Thus by (7)

$$\left(\int_{B_1} r^{2-n}\left|\frac{\partial \tilde{u}}{\partial r}\right|^2\right)^{2-\alpha} \leq C \int_{B_{3/2}\setminus B_{3/4}} r^{2-n}\left|\frac{\partial \tilde{u}}{\partial r}\right|^2,$$

provided $\|\tilde{u} - \varphi\|_{L^2(B_{3/2}\setminus B_{3/4})} < \eta$. By re-scaling, since $\frac{2}{3} \cdot \frac{3}{4} = \frac{1}{2}$, we in fact deduce

(8) $$\left(\int_{B_{1/2}} r^{2-n}\left|\frac{\partial \tilde{u}}{\partial r}\right|^2\right)^{2-\alpha} \leq C \int_{B_1 \setminus B_{1/2}} r^{2-n}\left|\frac{\partial \tilde{u}}{\partial r}\right|^2$$

for $\|u - \varphi\|^2_{L^2(B_1 \setminus B_{1/2})} < \eta^2$, where C is a constant depending only on φ. This is the key inequality; we claim that the theorem follows quite directly from it.

We need a simple inequality for real numbers, as follows:

(9)
$$(0 < a < b \leq 1, \alpha \in (0,1), \beta > 0, \text{ and } a^{2-\alpha} \leq \beta(b-a)) \Rightarrow a^{\alpha-1} - b^{\alpha-1} \geq C,$$

where C is a fixed constant determined by α, β only, and not depending on a, b. This is readily checked by calculus, considering separately the cases when $b/a \geq 2$ and $b/a < 2$. Notice that in view of (8) and (5) (with $\sigma\rho$ in place of ρ—notice that this amounts to applying the above discussion with $\tilde{u}^{(\sigma)}(x) \equiv \tilde{u}(\sigma x)$ in place of $\tilde{u}$), we can apply (9) with $a = I(\sigma/2)$ and $b = I(\sigma)$, where $I(\sigma) = \int_{B_s} r^{2-n}\left|\frac{\partial \tilde{u}}{\partial r}\right|^2$, thus giving

(10) $\quad I(\sigma/2)^{\alpha-1} - I(\sigma)^{\alpha-1} \geq C \quad$ provided $\sigma^{-n}\|\tilde{u} - \varphi\|^2_{L^2(B_\sigma \setminus B_{\sigma/2})} < \eta^2$,

provided $\sigma^{-n}\|\widetilde{u} - \varphi\|^2_{L^2(B_\sigma \setminus B_{\sigma/2})} < \eta^2$. Now $\int_{B_{\rho_2} \setminus B_{\rho_1}} |\widetilde{u} - \varphi|^2 = \int_{\rho_1}^{\rho_2} \tau^{n-1} \|u(\tau) - \varphi\|^2_{L^2(S^{n-1})}\,d\tau$, hence the condition $\sigma^{-n}\|\widetilde{u} - \varphi\|^2_{L^2(B_\sigma \setminus B_{\sigma/2})} < \eta^2$ (and hence the inequality in (10)) certainly holds if we require

(11) $$\|\widetilde{u}(\tau) - \varphi\|_{L^2(S^{n-1})} < \eta, \quad \forall \tau \in [\sigma/2, \sigma].$$

Now let us suppose that $\sigma \in (0, \frac{1}{2}]$ is given, take the unique integer $k \geq 1$ such that $\sigma \in [2^{-k-1}, 2^{-k})$, and assume

(12) $$\|\widetilde{u}(s) - \varphi\|_{L^2(S^{n-1})} < \eta, \quad \forall s \in [\sigma, 1].$$

Then we can apply (10) with $1/2^\ell$ in place of σ for $\ell = 0, \ldots, k$, whereupon we obtain by summing over $\ell = 0, \ldots, k$

$$I(2^{-k})^{\alpha-1} \geq I(2^{-k})^{\alpha-1} - I(1)^{\alpha-1} \geq Ck,$$

and hence

(13) $$I(\sigma) \equiv \int_0^\sigma r \left\|\frac{\partial \widetilde{u}}{\partial r}\right\|^2_{L^2(S^{n-1})} dr \leq \frac{C}{|\log \sigma|^{1+\beta}}, \quad \text{where } \beta = (1-\alpha)^{-1} - 1 > 0,$$

provided (12) holds. Notice integration by parts gives the general formula

$$\int_0^1 |\log r|^{1+\beta/2} r f(r)\,dr = \left. |\log r|^{1+\beta/2} \int_0^r s f(s)\,ds \right|_0^1 +$$
$$+ (1+\beta/2) \int_0^1 r^{-1} |\log r|^{\beta/2} \int_0^r s f(s)\,ds\,dr,$$

and using this with $f(r) = \|\frac{\partial \widetilde{u}}{\partial r}\|^2_{L^2(S^{n-1})}$ we obtain by virtue of (13) that

(14) $$\int_0^1 |\log r|^{1+\beta/2} r \left\|\frac{\partial \widetilde{u}}{\partial r}\right\|^2_{L^2(S^{n-1})} dr \leq C \int_0^1 \frac{1}{s|\log s|^{1+\beta/2}} \leq C,$$

again subject to (12). But then we have by virtue of Cauchy-Schwarz that

(15) $\|\widetilde{u}(\sigma) - \widetilde{u}(\tau)\|_{L^2(S^{n-1})}$

$$\leq \int_\sigma^\tau \left\|\frac{\partial \widetilde{u}(r)}{\partial r}\right\|_{L^2(S^{n-1})} dr$$
$$\leq \left(\int_\sigma^\tau r |\log r|^{1+\beta/2} \left\|\frac{\partial \widetilde{u}(r)}{\partial r}\right\|^2\right)^{1/2} \left(\int_\sigma^\tau r^{-1} |\log r|^{-1-\beta/2}\right)^{1/2}$$
$$\leq C|\log \tau|^{-\beta/2},$$

for any $0 < \sigma \leq \tau \leq \frac{1}{2}$, provided only that (12) holds. Next note that by another application of the Cauchy-Schwarz inequality we have

(16) $$\|\widetilde{u}(\tau) - \widetilde{u}(1)\|_{L^2(S^{n-1})} \leq \int_\tau^1 \left\|\frac{\partial \widetilde{u}}{\partial r}\right\|_{L^2(S^{n-1})} dr$$
$$\leq |\log \tau|^{1/2} \left(\int_{B_1} r^{2-n} \left|\frac{\partial \widetilde{u}}{\partial r}\right|^2\right)^{1/2} \equiv |\log \tau|^{1/2} \varepsilon,$$

where $\varepsilon = (\int_{B_1} r^{2-n} |\frac{\partial \widetilde{u}}{\partial r}|^2)^{1/2} = (\int_{B_\rho(y)} r^{2-n} |\frac{\partial u}{\partial r}|^2)^{1/2}$ ($\to 0$ as $\rho \downarrow 0$). Now notice that, by virtue of the triangle inequality, (16) already guarantees that

$$\|\widetilde{u}(\tau) - \varphi\|_{L^2(S^{n-1})} < \eta/2, \quad \forall \sigma \leq \tau \leq 1$$

if

(17) $\qquad \varepsilon |\log \sigma|^{1/2} < \eta/4$ and $\|\widetilde{u}(1) - \varphi\|_{L^2(S^{n-1})} < \eta/4$.

So now suppose (17) holds and choose $\tau \in (0, 1/2)$ such that $C|\log \tau|^{-\beta/2} < \eta/4$. Then $\|\widetilde{u}(\sigma) - \varphi\|_{L^2} \leq \|\widetilde{u}(\sigma) - \widetilde{u}(\tau)\|_{L^2} + \|\widetilde{u}(\tau) - \varphi\|_{L^2}$ and hence by (15), (17) we deduce that $\|\widetilde{u}(\sigma) - \varphi\|_{L^2(S^{n-1})} < \eta/2$ so long as $\|\widetilde{u}(s) - \varphi\|_{L^2(S^{n-1})} < \eta$ for $s \in [\sigma, 1]$. Clearly this shows that $\|\widetilde{u}(\sigma) - \varphi\|_{L^2(S^{n-1})} < \eta/2$ for all $\sigma \in (0, 1]$ provided only we can ensure that ε can be selected so that (17) holds. However, $\widetilde{u} = u_{y,\rho}$, so by taking $\rho = \rho_j$ with j sufficiently large, where $\rho_j \downarrow 0$ is such that $u_{y,\rho_j} \to \varphi$, we can of course ensure both inequalities in (17). (The second we already discussed at the beginning of the proof and the first trivially holds for j sufficiently large because $\int_{B_\rho} r^{2-n} |\frac{\partial u}{\partial r}|^2 \to 0$ as $\rho \downarrow 0$.) But now this means that (12) holds for all $\sigma \in (0, 1]$ and hence we can apply (15) with any σ, τ. Then letting $\sigma = \sigma_j$ such that $\widetilde{u}(\sigma_j) \to \varphi$ (which we can do because φ is a tangent map of $\widetilde{u}$ at 0), we then have

$$\|\widetilde{u}(\tau) - \varphi\|_{L^2(S^{n-1})} \leq C|\log \tau|^{-\beta/2}, \quad \tau \in (0, \frac{1}{2}],$$

which is the required asymptotic decay. $\square$

3.16 Appendix to Chapter 3

3.16.1 The Liapunov-Schmidt Reduction in a Finite Dimensional Setting

Suppose $F \in C^\infty(\Omega)$ where Ω is an open subset of some Euclidean space $\mathbb{R}^Q$ with $0 \in \Omega$. Suppose that 0 is a critical point of F; that is,

(i) $\qquad\qquad\qquad \nabla F(0) = 0,$

and suppose (without loss of generality in view of the possibility of composing F with an orthogonal transformation) that

(ii) $\qquad\qquad\qquad \ker \operatorname{Hess} F|_0 = \mathbb{R}^q \times \{0\},$

where $0 \leq q \leq Q$ and where $\operatorname{Hess} F|_0$ (the Hessian of F at 0) is viewed as a symmetric linear transformation of $\mathbb{R}^Q$. Then we have the following lemma, which effectively reduces the study of the critical points of F near 0 to the study of the critical points of a related function f defined in a neighbourhood of 0 in the lower dimensional space $\mathbb{R}^q \times \{0\}$.

Lemma 1 *If $q \in \{0, \ldots, Q-1\}$ and if (i), (ii) hold, then there is a neighbourhood W of 0 in Ω such that there is a diffeomorphism Ψ of W onto a neighbourhood U of 0 in $\mathbb{R}^Q$ with*

$$\{x \in U \cap P^{-1}(W \cap (\mathbb{R}^q \times \{0\}))\ :\ \nabla F(x) = 0\} = \Psi\{\xi \in W \cap (\mathbb{R}^q \times \{0\})\ :\ \nabla f(\xi) = 0\},$$

where
$$f(\xi) = F(\Psi(\xi)), \quad \xi \in W \cap (\mathbb{R}^q \times \{0\})$$

and P denotes orthogonal projection of x onto $\mathbb{R}^q \times \{0\}$; Ψ is in fact explicitly given as the local inverse of the map $x \mapsto x' + \nabla F(x)$ in some neighbourhood of the origin, where $x' = P(x)$. Furthermore W can be chosen so that we have the inequalities

$$C^{-1}|\nabla F(\Psi(x'))| \le |\nabla f(x')| \le C|\nabla F(x)|, \qquad |f(x') - F(x)| \le C|\nabla F(x)|^2$$

for every $x \in U$ such that $x' \in W \cap (\mathbb{R}^q \times \{0\})$.

Remark: If $q = 0$ then $\operatorname{Hess} F|_0$ is an isomorphism of $\mathbb{R}^Q$, and hence the inverse function theorem tells us that $x \mapsto \nabla F(x)$ is a diffeomorphism of suitable neighbourhoods of 0, so the lemma holds trivially in this case (interpreting $\nabla f \equiv 0$, which is reasonable since $f(0) = 0$ and f is only defined on the zero-dimensional subspace $\{0\}$).

Proof: Let

$$\mathcal{N}(x) = x' + \nabla F(x), \quad x \in \Omega, \tag{1}$$

and note that

$$d\mathcal{N}|_0(v) = v' + \operatorname{Hess} F|_0(v), \quad v \in \mathbb{R}^Q.$$

In particular (since $\operatorname{Hess} F|_0$ is injective on $\{0\} \times \mathbb{R}^{Q-q} = \operatorname{range} \operatorname{Hess} F|_0$) we have $d\mathcal{N}|_0(v) = 0$ if and only if both $v' = 0$ and $\operatorname{Hess} F|_0 v = 0$ so that $v \in (\{0\} \times \mathbb{R}^{Q-q}) \cap (\mathbb{R}^q \times \{0\}) = \{0\}$, i.e. $v = 0$. Thus $d\mathcal{N}|_0$ is injective and hence by the inverse function theorem there are bounded neighbourhoods U_0, V_0 of 0 with smooth boundaries such that $\mathcal{N}|\overline{U}_0$ is a diffeomorphism of $\overline{U}_0$ onto $\overline{V}_0$ with smooth inverse Ψ on $\overline{V}_0$. Since $\mathcal{N}(\Psi(x)) \equiv x$, $x \in V_0$, we have in particular that

$$\begin{aligned}(\nabla F(\Psi(x')))^\perp &= (\mathcal{N}(\Psi(x')) - (\Psi(x'))')^\perp \\ &= (x' - (\Psi(x'))')^\perp \equiv 0, \quad x' \in (\mathbb{R}^q \times \{0\}) \cap V_0,\end{aligned} \tag{2}$$

where $(\cdot)^\perp$ denotes orthogonal projection into $\{0\} \times \mathbb{R}^{Q-q}$. By 1-dimensional calculus along the line segment joining z to $\mathcal{N}(y)$ we have

$$\begin{aligned}|\Psi(z) - y| &\equiv |\Psi(z) - \Psi(\mathcal{N}(y))| \\ &= \left|\int_0^1 \langle \nabla \Psi(z + s(\mathcal{N}(y) - z)), \mathcal{N}(y) - z \rangle\, ds\right| \\ &\le C|z - \mathcal{N}(y)|\end{aligned} \tag{3}$$

3.16. Appendix to Chapter 3

for $z \in B_\sigma$ and $y \in U$, where $B_\sigma = B_\sigma(0) \subset V_0$ and $U = \Psi(B_\sigma)$. With $z = y'$, this gives

(4) $\qquad |\Psi(y') - y| \leq C|\nabla F(y)|, \quad y \in \tilde{U}_0, \; y' \in (\mathbb{R}^q \times \{0\}) \cap \tilde{V}_0$

by virtue of the fact that $y' - \mathcal{N}(y) = -\nabla F(y)$ by definition of $\mathcal{N}$.

Now we define
$$f(\xi) = F(\Psi(\xi)), \quad \xi \in B_\sigma \cap (\mathbb{R}^q \times \{0\})$$
and note that by the chain rule

(5) $\qquad \langle v, \nabla f(\xi) \rangle = \langle \nabla F(\Psi(\xi)), d\Psi|_\xi(v) \rangle, \quad \xi \in W_0, \; v \in \mathbb{R}^q \times \{0\},$

and hence, for $y \in U$ such that $y' \in B_\sigma \cap (\mathbb{R}^q \times \{0\})$,

(6) $\qquad \begin{aligned} |\nabla f(y')| &\leq C|\nabla F(\Psi(y'))| \\ &= C|\nabla F(y + \Psi(y') - y)| \\ &\leq C|\nabla F(y + \Psi(y') - y) - \nabla F(y)| + C|\nabla F(y)| \\ &\leq C|\Psi(y') - y| + C|\nabla F(y)| \\ &\leq C|\nabla F(y)|, \end{aligned}$

where we used (4) again. Using (4) yet again, this time with y' in place of y, we have $|\Psi(y') - y'| \leq C|\nabla F(y')|$, and since $\nabla F(0) = 0$ and $\operatorname{Hess} F|_0(\mathbb{R}^q \times \{0\}) = \{0\}$, this implies that $|\Psi(y') - y'| \leq C|y'|^2$ for $y \in U$ such that $y' \in B_\sigma \cap (\mathbb{R}^q \times \{0\})$, which in particular implies that

(7) $\qquad d(\Psi \circ P)|_0 = P,$

where P is orthogonal projection onto $\mathbb{R}^q \times \{0\}$. But now, for $\xi \in B_\sigma \cap (\mathbb{R}^q \times \{0\})$, by (5) we have

$$\langle v, \nabla f(\xi) \rangle = \langle \nabla F(\Psi(\xi)), d\Psi|_\xi(v) \rangle = $$
$$= \langle \nabla F(\Psi(\xi)), v \rangle + \langle \nabla F(\Psi(\xi)), (d(\Psi \circ P)|_\xi - P)v \rangle, \quad v \in \mathbb{R}^q \times \{0\},$$

and $|(d(\Psi \circ P)|_\xi - P)| \leq C|\xi|$ by (7). Thus, in case $\nabla F(\Psi(\xi)) \neq 0$, since $\nabla F(\Psi(\xi)) \in \mathbb{R}^q \times \{0\}$ by (2), we can first choose v parallel to $\nabla F(\Psi(\xi))$ in order to conclude

$$|\nabla F(\Psi(\xi))| \leq |\nabla f(\xi)| + C|\xi| \, |\nabla F(\Psi(\xi))|,$$

and then choose η parallel to $\nabla f(\xi)$ in order to conclude

$$|\nabla F(\Psi(\xi))| \geq |\nabla f(\xi)| - C|\xi| \, |\nabla F(\Psi(\xi))|.$$

Hence we conclude

(8) $\qquad \dfrac{1}{2}|\nabla f(\xi)| \leq |\nabla F(\Psi(\xi))| \leq 2|\nabla f(\xi)|$

for $\xi \in B_\sigma \cap (\mathbb{R}^q \times \{0\})$, provided σ is sufficiently small.

Next we note that for $y \in U$ such that $y' \in B_\sigma \cap (\mathbb{R}^q \times \{0\})$

$$\nabla F(y) = 0 \Rightarrow \mathcal{N}(y) = y' \Rightarrow y = \Psi(y'),$$

so that by (8) we have

$$\nabla F(y) = 0 \Rightarrow y = \Psi(y') \text{ and } \nabla f(y') = 0.$$

Conversely if $y = \Psi(\xi)$ with $\xi \in B_\sigma \cap (\mathbb{R}^q \times 0)$ and if $\nabla f(\xi) = 0$ then $\nabla F(y) = 0$ by (8), and then also $y' = \mathcal{N}\Psi(\xi) = \xi$. Thus we have

$$\{y \in U \cap P^{-1}(B_\sigma \cap (\mathbb{R}^q \times \{0\})) : \nabla F(y) = 0\} = \Psi\{\xi \in B_\sigma \cap (\mathbb{R}^q \times 0) : \nabla f(\xi) = 0\},$$

where P denotes orthogonal projection onto $\mathbb{R}^q \times \{0\}$.

This completes the proof of the lemma (with $W = B_\sigma$), except for the second inequality in the last part of the lemma. To prove this we note that for $y \in U \cap P^{-1}(W \cap (\mathbb{R}^q \times \{0\}))$

$$\begin{aligned}
|f(y') - F(y)| &= |F(\Psi(y')) - F(y)| \\
&= \left| \int_0^1 \langle \nabla F(y + s(\Psi(y') - y)), \Psi(y') - y \rangle \, ds \right| \\
&\leq \int_0^1 |\nabla F(y + s(\Psi(y') - y))| \, ds \, |\Psi(y') - y| \\
&\leq \left(|\nabla F(y)| + \int_0^1 |\nabla F(y + s(\Psi(y') - y)) - \nabla F(y)| \right) |\Psi(y') - y| \\
&\leq C|\nabla F(y)|^2
\end{aligned}$$

by (4) again. $\square$

Chapter 4

Rectifiability of the Singular Set

In this chapter we establish rectifiability results for the singular set sing u of energy minimizing maps.

4.1 Statement of Main Theorems

Recall that a subset $A \subset \mathbb{R}^n$ is said to be m-rectifiable if $\mathcal{H}^m(A) < \infty$, and if A has an approximate tangent space a.e. in the sense that for $\mathcal{H}^m$-a.e. $z \in A$ there is an m-dimensional subspace L_z such that

$$\lim_{\sigma \downarrow 0} \int_{\eta_{z,\sigma}(A)} f \, d\mathcal{H}^m = \int_{L_z} f \, d\mathcal{H}^m, \quad f \in C_c^0(\mathbb{R}^n),$$

where, here and subsequently, $\eta_{z,\sigma}(x) \equiv \sigma^{-1}(x-z)$.

A subset $A \subset \mathbb{R}^n$ is said to be locally m-rectifiable if it is m-rectifiable in a neighbourhood of each of its points. Thus for each $z \in A$ there is a $\sigma > 0$ such that $A \cap B_\sigma(z)$ is m-rectifiable.

Similarly A is locally compact if for each $z \in A$ there is $\sigma > 0$ such that $A \cap \overline{B}_\sigma(z)$ is compact.

This terminology is used in the statement of the main theorems below.

Our main theorem for the case when the target manifold is real-analytic is the following:

Theorem 1 *If $u \in W^{1,2}(\Omega; N)$ is energy minimizing and if N is real-analytic, then, for each closed ball $B \subset \Omega$, sing $u \cap B$ is the union of a finite pairwise disjoint collection of locally $(n-3)$-rectifiable locally compact subsets.*

Remarks: (1) Notice that being a finite union of locally m-rectifiable subsets is slightly weaker than being a (single) locally m-rectifiable subset, in that if $A = \cup_{k=1}^{Q} A_k$, where each A_k is locally m-rectifiable, there may be a set of points y of positive measure on one of the A_ℓ such that $\mathcal{H}^m((\cup_{k\neq \ell} A_k) \cap B_\sigma(y)) = \infty$ for each $\sigma > 0$. (This is possible because A_k has locally finite measure in a neighbourhood of each of its points, but may not have locally finite measure in a neighbourhood of points in the closure $\overline{A}_k$ and this may intersect A_ℓ, $\ell \neq k$.)

(2) We shall prove also that $\Theta_u(z)$ is a.e. constant on each of the sets in the finite collection referred to in the above theorem, and furthermore it will be established that sing u has a (unique) tangent space in the Hausdorff distance sense at $\mathcal{H}^{n-3}$-almost all points $x \in \operatorname{sing} u$.

For the case when N is merely smooth rather than real analytic, we need to assume an "integrability condition" as in (xiii) of Section 3.14. Thus we have:

Theorem 2 *The conclusions of Theorem 1 continue to hold if the requirement that N is real analytic is dropped (so N is an arbitrary smooth compact Riemannian manifold isometrically embedded in $\mathbb{R}^P$), provided the integrability condition (xiii) of Section 3.14 holds for all smooth harmonic maps $\varphi_0 : S^2 \to N$ which are such that the homogeneous degree zero extension of φ_0 to $\mathbb{R}^3 \setminus \{0\}$ is a locally energy minimizing map of $\mathbb{R}^3$ into N.*

Remarks: (1) The stronger conclusions of remark (2) above also hold in this case.

(2) If N is the standard S^2 or $\mathbb{R}P^2$, or metrically sufficiently close to the standard S^2 or $\mathbb{R}P^2$ in the C^3 sense, then the following much stronger conclusion holds:

$$\operatorname{sing} u = \left(\bigcup_{j=1}^{Q} \Sigma_j\right) \cup K$$

where each Σ_j is a properly embedded $C^{1,\alpha}$ manifold and K is a closed set of dimension $\leq n - 4$. This is proved, by rather different techniques ("blowing up") than those used here. For a detailed discussion of the blowing up method in the context of nonisolated singularities of minimal surfaces we refer to [Si93], and for an outline of the proof (using the same blowup method) of the above result for energy minimizing maps we refer to [Si92]; the detailed proof, which very precisely parallels the corresponding result for minimal surfaces proved in [Si93], will appear in [Si].

4.2 A general rectifiability lemma

Here we develop a general rectifiability lemma, which gives sufficient conditions for an arbitrary closed subset of $\mathbb{R}^n$ to be m-rectifiable, as defined in Section 4.1. This rectifiability lemma will be crucial in our later rectifiability proofs of Section 4.7, where it will be applied with $m = n - 3$.

4.2. A general rectifiability lemma

Let $S \subset \mathbb{R}^n$ be an arbitrary closed set, $\varepsilon, \delta \in (0,1)$ with $\varepsilon < \frac{\delta}{16}$ (in the applications below we always have $\varepsilon \ll \delta$), $\rho_0 > 0$, and assume S has the ε-approximation property satisfied for S_+ in Lemma 2 of Section 3.9. Thus for each $y \in S$ and each $\rho \in (0, \rho_0]$ we assume

(i) $\qquad S \cap B_\rho(y) \subset$ the $(\varepsilon \rho)$-neighbourhood
of some m-dimensional affine space $L_{y,\rho}$ containing y.

In all that follows we assume that such $L_{y,\rho}$, corresponding to each $y \in S$ and $\rho \leq \rho_0$, are fixed. Then, relative to such a choice, we have the following definition.

Definition 1 *With the notation in (i) above, we say S has a δ-gap in a ball $B_\rho(y)$ with $y \in S$ if there is $z \in L_{y,\rho} \cap \overline{B}_{(1-\delta)\rho}(y)$ such that $B_{\delta\rho}(z) \cap S = \varnothing$.*

With this terminology and for any given $y \in S$, $\rho \in (0, \rho_0]$ and $\delta \in (0, \frac{1}{2})$, we define

(ii) $\qquad \gamma(y, \rho, \delta) = \sup(\{0\} \cup \{\sigma \in (0, \rho] : S \text{ has a } \delta\text{-gap in } B_\sigma(y)\})$.

Thus $\gamma(y, \rho, \delta) = 0$ means that S has no δ-gaps in the balls $B_\tau(y)$, $\tau \in (0, \rho]$, and if $\gamma(y, \rho, \delta) > 0$ then S has no δ-gaps in the balls $B_\tau(y)$, $\tau \in (\gamma(y, \rho, \delta), \rho]$, but S does have a δ-gap in $B_{\tau_j}(y)$ for some sequence $\tau_j \uparrow \gamma(y, \rho, \delta)$; in particular, S has a $\frac{\delta}{2}$-gap in $B_{\gamma(y,\rho,\delta)}(y)$ in this case.

Lemma 1 (Rectifiability Lemma) *Let $\delta \in (0, \frac{1}{32})$ be given. There exists $\varepsilon = \varepsilon(m, n, \delta) \in (0, \frac{\delta}{16})$ such that the following holds. Let $\rho_0 > 0$, $x_0 \in S = $ a closed subset of $\mathbb{R}^n$ satisfying the ε-approximation property (i), and suppose:*

(I) *Either S has a $\frac{\delta}{20}$-gap in $B_{\rho_0}(x_0)$ or there is an m-dimensional subspace $L_0 \subset \mathbb{R}^n$ and a family $\mathcal{F}_0$ of balls with centers in $S \cap \overline{B}_{\rho_0}(x_0)$ such that*

(a) $\qquad \sum_{B \in \mathcal{F}_0} (\operatorname{diam} B)^m \leq \varepsilon \rho_0^m$,

(b) $\qquad S \cap B_\sigma(y) \subset$ the $(\varepsilon \sigma)$-neighbourhood of $y + L_0$,
$\qquad\qquad \forall\, y \in S \cap B_{\rho_0/2}(x_0) \setminus (\cup \mathcal{F}_0), \ \sigma \in (\gamma(y, \frac{\rho_0}{2}, \delta), \frac{\rho_0}{2}]$

(with $\gamma(y, \rho, \delta)$ as in (ii)), and

(II) *$\forall x_1 \in S \cap \overline{B}_{\rho_0}(x_0)$ and $\forall \rho_1 \in (0, \frac{\rho_0}{2}]$ there are $L_1, \mathcal{F}_1$ (depending on x_1, ρ_1) such that the hypotheses (I) continue to hold with $x_1, \rho_1, L_1, \mathcal{F}_1$ in place of $x_0, \rho_0, L_0, \mathcal{F}_0$ respectively.*

Then $S \cap \overline{B}_{\rho_0}(x_0)$ is m-rectifiable.

Remarks: (1) It is important, from the point of view of the application which we have in mind, that the property (I)(b) need only be checked on balls $B_\sigma(y)$ such that S has no δ-gap in any of the balls $B_\tau(x_0)$, $\gamma(y, \frac{\rho_0}{2}, \delta) < \tau \leq \frac{\rho_0}{2}$. In practice the condition (II) is often an automatic consequence of the way in which S is defined in the first place. For example, in Section 4.7 below we apply the above lemma with $x_0 = 0$ and $\rho_0 = 1$ to a subset S_+ of the singular set satisfying certain hypotheses ((i)–(v) of Section 4.5) which are automatically satisfied with $\widetilde{S}_+ = \eta_{x_1,\rho_1} S_+$ in place of S_+ and $\tilde{u} = u \circ \eta_{x_1,\rho_1}$ in place of u, where $\eta_{x_1,\rho_1} : x \mapsto \rho_1^{-1}(x - x_1)$. Thus once we have checked that the hypotheses (i)–(v) of Section 4.5 imply that (I) holds for S_+, then it *automatically* follows that the hypotheses (II) also hold for S_+.

(2) Notice that if S does not have a $\frac{\delta}{20}$-gap in $B_{\rho_0}(x_0)$ (so that the first alternative hypothesis of (I) does *not* hold), then, provided ε is sufficiently small relative to δ, no ball $B_\tau(y)$ for $\tau \in [\frac{\rho_0}{16}, \frac{\rho_0}{2}]$ and $y \in S \cap B_{\rho_0/2}(x_0)$ can have a δ-gap, so in particular $\gamma(y, \frac{\rho_0}{2}, \delta) \leq \frac{\rho_0}{16}$ and condition I(b) always has non-trivial content in this case.

(3) In order to establish the Theorems 1, 2 of Section 4.1 we are going to show that this lemma can be applied with sets S of the form $S_+ = \overline{B}_\rho(y) \cap \{x \in \text{sing } u : \Theta_u(x) \geq \Theta_u(y)\}$ with suitable $y \in \text{sing } u$ and with ρ sufficiently small. Notice that Lemma 2 of Section 3.9 already establishes the weak ε-approximation property (i) for such S_+. Most importantly, we are able in the discussion of Section 4.5 to get much more control on $\text{sing } u$ in balls which do not have δ-gaps. This is the key point which makes it possible to check the hypotheses I and hence to prove the main theorems stated in the introduction.

In the proof of the rectifiability lemma, we shall need the following covering lemma.

Lemma 2 *If $\delta \in (0, \frac{1}{32})$, $F \subset \overline{B}_{\rho_0}(x_0) \subset \mathbb{R}^m$ is arbitrary, and if $\mathcal{B}$ is a collection of closed balls of radius $\leq \frac{\rho_0}{8}$ and centers in F which covers F, and for each $B = \overline{B}_\rho(y) \in \mathcal{B}$ there is $z \in \overline{B}_{(1-\delta)\rho}(y)$ such that $B_{\delta\rho}(z) \cap F = \emptyset$ (that is, F has a δ-gap in each ball $B \in \mathcal{B}$), then there is a covering $\mathcal{U} = \{B_{\rho_k}(y_k)\}$ of F by balls with centers $y_k \in F$ and with*

$$\sum_k \rho_k^m \leq (1-\theta)\rho_0^m, \quad \theta = \theta(\delta, m) \in (0,1),$$

and, also, for each $B_{\rho_k}(y_k) \in \mathcal{U}$ there is a ball $\overline{B}_{\sigma_k}(z_k) \in \mathcal{B}$ such that $B_{\theta^{-1}\rho_k}(y_k) \supset B_{\sigma_k}(z_k)$.

Proof: By translation and a scaling, we can assume $\rho_0 = 1$, $x_0 = 0$. We here consider closed cubes $Q = [z_1 - r, z_1 + r] \times \cdots \times [z_m - r, z_m + r]$; $z = (z_1, \ldots, z_m)$ is the center of Q and $2r$ is the edge-length, denoted $e(Q)$. For integers $N \geq 2$, $Q^{(N)}$ will denote the N-times enlargement of Q; thus $Q^{(N)} = [z_1 - Nr, z_1 + Nr] \times \cdots \times [z_m - Nr, z_m + Nr]$.

We first construct a cover $\mathcal{U}_0 = \{B_{\rho_k}(y_k)\}$ for F which satisfies all the stated conditions, except possibly for the requirement that each $y_k \in F$.

As an initial observation, we note that such a collection $\mathcal{U}_0$ trivially exists with $\theta = \theta(m) \in (0,1)$ (independent of δ) if $F \cap [-\frac{1}{4\sqrt{m}}, \frac{1}{4\sqrt{m}}]^m = \emptyset$. Similarly, for any

4.2. A general rectifiability lemma

given $\sigma_0 \in (0, \frac{1}{8})$ it is easy to check the existence of such $\mathcal{U}_0$ with $\theta = \theta(m, \delta, \sigma_0) \in (0, 1)$ if there is a ball $B \in \mathcal{B}$ such that radius $B \geq \sigma_0$ and $B \cap [-\frac{1}{2\sqrt{m}}, \frac{1}{2\sqrt{m}}]^m \neq \emptyset$, because then, since radius $B \leq \frac{1}{8}$, we have $B \subset B_1(0)$ and hence there is a ball $B_{\delta\sigma_0}(z) \subset B_1(0)$ with $F \cap B_{\delta\sigma_0}(z) = \emptyset$.

So subsequently we can assume without loss of generality that

(1) $$F \cap [-\tfrac{1}{4\sqrt{m}}, \tfrac{1}{4\sqrt{m}}]^m \neq \emptyset,$$

and that

(2) $$B \cap Q_0 \neq \emptyset \Rightarrow \operatorname{diam} B \leq \tfrac{1}{48\sqrt{m}},$$

where $Q_0 = [-\frac{1}{\sqrt{m}}, \frac{1}{4\sqrt{m}}]^m \subset \overline{B}_1(0)$. Define

(3) $$F_0 = F \cap Q_0, \quad \mathcal{B}_0 = \{B \in \mathcal{B} : B \cap Q_0 \neq \emptyset\},$$

so that $\operatorname{diam} B \leq \frac{1}{48\sqrt{m}}$ for each $B \in \mathcal{B}_0$ by (2). For each $k = 1, 2, \ldots$ let $\mathcal{C}_k$ be the collection of $2^{(k+1)m}$ congruent subcubes of edge-length $\frac{1}{\sqrt{m}} 2^{-k}$ obtained by repeated subdivision ($k+1$ times) of the cube Q_0, and let $\mathcal{C} = \cup_{k \geq 1} \mathcal{C}_k$. Now for each $B \in \mathcal{B}_0$ let k_B be the unique positive integer such that

$$\operatorname{diam} B < \tfrac{2^{-k_B}}{\sqrt{m}} \leq 2 \operatorname{diam} B.$$

(Since $\operatorname{diam} B \leq \frac{1}{48\sqrt{m}}$, we actually must have $k_B \geq 5$.)

For each $B \in \mathcal{B}_0$ let $\mathcal{Q}_B$ be the collection of $Q \in \mathcal{C}_{k_B}$ such that $Q \cap B \neq \emptyset$. Notice that

(4) $$B \subset Q^{(3)} \text{ for each } Q \in \mathcal{Q}_B,$$

because $e(Q) > \operatorname{diam} B$ and $B \cap P \neq \emptyset$ for each $P, Q \in \mathcal{Q}_B$. Also, since $\operatorname{diam} B \leq \frac{1}{48\sqrt{m}}$ for each $B \in \mathcal{B}_0$, we have

(5) $$e(Q) \leq 2 \operatorname{diam} B \leq \tfrac{1}{24\sqrt{m}} \quad \forall B \in \mathcal{B}_0, \, Q \in \mathcal{Q}_B.$$

Now for $k = 1, 2, \ldots$, we define

(6) $$\mathcal{Q}_k = \bigcup_{B \in \mathcal{B}_0, \, k_B = k} \mathcal{Q}_B,$$

and

$$\mathcal{Q} = \bigcup_k \mathcal{Q}_k.$$

Notice that then

(7) $$\bigcup_{Q \in \mathcal{Q}} Q \supset Q_0 \cap \left(\bigcup_{B \in \mathcal{B}_0} B \right) \supset F_0.$$

Notice also that by (4) and (5) we have

(8) $\quad Q \in \mathcal{Q}$ and $Q \subset [-\frac{1}{2\sqrt{m}}, \frac{1}{2\sqrt{m}}]^m \Rightarrow$
$$\exists\, B \in \mathcal{B}_0 \text{ with } Q \in \mathcal{Q}_B \text{ and } B \subset \bigcup_{P \in \mathcal{Q}_B} P \subset Q^{(3)} \subset Q_0.$$

In particular, since F_0 has a δ-gap in any such B, we have

(9) $\quad Q \in \mathcal{Q}$ and $Q \subset [-\frac{1}{2\sqrt{m}}, \frac{1}{2\sqrt{m}}]^m \Rightarrow$
$$\exists\, Q_1 \in \mathcal{Q} \text{ with } e(Q_1) = e(Q),\ R \subset Q_1 \subset Q^{(3)} \subset Q_0,$$

where R is a cube satisfying

(10) $\qquad e(R) \geq \frac{\delta}{4\sqrt{m}} e(Q) \text{ and } R \cap F_0 = \varnothing.$

Now we define a subcollection $\widetilde{\mathcal{Q}}$ of $\mathcal{Q}$ as follows:

First, let $\widetilde{\mathcal{Q}}_1 = \mathcal{Q}_1$, and for each $k \geq 1$ (assuming $\widetilde{\mathcal{Q}}_1, \ldots, \widetilde{\mathcal{Q}}_k$ are already defined) let

$$\widetilde{\mathcal{Q}}_{k+1} = \left\{ Q \in \mathcal{Q}_{k+1} : Q \not\subset \bigcup_{\ell=1}^{k} \bigcup_{P \in \widetilde{\mathcal{Q}}_\ell} P \right\},$$

and then define $\widetilde{\mathcal{Q}} = \cup_k \widetilde{\mathcal{Q}}_k$. Evidently $\widetilde{\mathcal{Q}}_k \subset \mathcal{Q}_k$, $\cup \widetilde{\mathcal{Q}} = \cup \mathcal{Q}$,

$$\{\text{interior}(Q) : Q \in \widetilde{\mathcal{Q}}\} \text{ is a pairwise disjoint collection.}$$

If $k \geq 1$ and if $Q \in \widetilde{\mathcal{Q}}_k$ with $Q \subset [-\frac{1}{2\sqrt{m}}, \frac{1}{2\sqrt{m}}]^m$, then either all $P \in \mathcal{Q}_k$ with $Q \cap P \neq \varnothing$ are in $\widetilde{\mathcal{Q}}_k$, in which case, by (9), (10),

(11) $\quad \exists\, Q_1 \in \widetilde{\mathcal{Q}}_k$ such that $Q_1 \subset Q^{(3)}$ and such that
$$\exists \text{ a cube } R \subset Q_1 \text{ with } F_0 \cap R = \varnothing \text{ and } e(R) \geq \frac{\delta}{4\sqrt{m}} e(Q_1),$$

or else there is $P_1 \in \mathcal{Q}_k$ with $Q \cap P_1 \neq \varnothing$ and $P_1 \notin \widetilde{\mathcal{Q}}_k$, in which case by definition of $\widetilde{\mathcal{Q}}_k$ we must have $P_1 \subset P$ for some $P \in \widetilde{\mathcal{Q}}_\ell$ with $\ell \leq k-1$ and hence $Q \subset P_1^{(3)} \subset P^{(3)}$. Notice that, in the latter case, since $e(P) \geq 2e(Q)$ and $Q \subset P^{(3)}$, we have $Q^{(6)} \subset P^{(6)}$. Proceeding inductively, starting with an arbitrary $Q \in \widetilde{\mathcal{Q}}$ with $Q \subset [-\frac{1}{4\sqrt{m}}, \frac{1}{4\sqrt{m}}]^m$, we then either have (11) or else there are integers $q \geq 1$ and $k-1 \geq \ell_1 > \cdots > \ell_q \geq 1$, and cubes $P_{\ell_j} \in \widetilde{\mathcal{Q}}_{\ell_j}$, $j = 1, \ldots, q$, such that

(12) $\qquad Q^{(6)} \subset P_{\ell_1}^{(6)} \subset \cdots \subset P_{\ell_q}^{(6)} \subset [-\frac{1}{2\sqrt{m}}, \frac{1}{2\sqrt{m}}]^m,$

$\exists Q_1 \in \widetilde{\mathcal{Q}}_{\ell_q}$ and a cube $R \subset Q_1 \subset P_{\ell_q}^{(3)}$ with $F_0 \cap R = \varnothing$ and $e(R) \geq \frac{\delta}{4\sqrt{m}} e(Q_1)$.

4.2. A general rectifiability lemma

(We can inductively establish the inclusion $P_{\ell_j}^{(6)} \subset [-\frac{1}{2\sqrt{m}}, \frac{1}{2\sqrt{m}}]^m$ by using the facts that $P_{\ell_j}^{(6)} \cap [-\frac{1}{4\sqrt{m}}, \frac{1}{4\sqrt{m}}]^m \supset P_{\ell_j}^{(6)} \cap Q \neq \varnothing$, and $e(P_{\ell_j}^{(6)}) \leq 6e(P_{\ell_j}) \leq \frac{1}{4\sqrt{m}}$ by (5).) Therefore if we let $\mathcal{G} \subset \tilde{\mathcal{Q}}$ be defined by

$$\mathcal{G} = \{Q_1 \in \tilde{\mathcal{Q}} : F_0 \cap R = \varnothing \text{ for some cube } R \subset Q_1 \text{ with } e(R) \geq \frac{\delta}{4\sqrt{m}} e(Q_1)\}$$

then, keeping in mind that $Q_1 \subset P^{(3)} \Rightarrow P^{(6)} \subset Q_1^{(8)}$ if $e(P) = e(Q_1)$, we have by (11) and (12) that

(13) $$\bigcup_{Q_1 \in \mathcal{G}} Q_1^{(8)} \supset \bigcup_{Q \in \tilde{\mathcal{Q}}, Q \subset [-\frac{1}{4\sqrt{m}}, \frac{1}{4\sqrt{m}}]^m} Q^{(6)} \supset \bigcup_{Q \in \tilde{\mathcal{Q}}, Q \subset [-\frac{1}{4\sqrt{m}}, \frac{1}{4\sqrt{m}}]^m} Q.$$

On the other hand we know by definition of $\mathcal{G}$ we can decompose each $Q \in \mathcal{G}$ into a union of a collection $\mathcal{P}_Q$ of congruent sub-cubes (with non-overlapping interiors), such that

(14) $$\frac{\delta}{32\sqrt{m}} e(Q) \leq e(P) \leq \frac{\delta}{8\sqrt{m}} e(Q) \quad \forall P \in \mathcal{P}_Q, \text{ and } F_0 \cap P = \varnothing \text{ for some } P \in \mathcal{P}_Q.$$

Then if we let $\mathcal{P}_Q^0$ be the subcollection of $\mathcal{P}_Q$ obtained by dropping all $P \in \mathcal{P}_Q$ such that $F_0 \cap P = \varnothing$, we have

(15) $$\sum_{P \in \mathcal{P}_Q^0} |P| \leq (1 - \theta \delta^m) |Q|, \quad Q \in \mathcal{G},$$

for suitable $\theta = \theta(m)$, where $|Q|$ denotes the volume of Q (that is, $e(Q)^m$). But by (13) we have

(16) $$\sum_{Q \in \mathcal{G}} |Q| \geq 8^{-m} \sum_{Q \in \tilde{\mathcal{Q}}, Q \subset [-\frac{1}{4\sqrt{m}}, \frac{1}{4\sqrt{m}}]^m} |Q|.$$

Now if

(17) $$\sum_{Q \in \tilde{\mathcal{Q}}, Q \subset [-\frac{1}{4\sqrt{m}}, \frac{1}{4\sqrt{m}}]^m} |Q| \leq \frac{1}{2} \frac{1}{(2\sqrt{m})^m}$$

then we have

(18) $$\sum_{Q \in \tilde{\mathcal{Q}}} |Q| \leq \frac{1}{2} \frac{1}{(2\sqrt{m})^m} + |Q_0| - \frac{1}{(2\sqrt{m})^m} \leq (1 - \frac{1}{4^{m+1}} \frac{1}{(2\sqrt{m})^m}) |Q_0|.$$

On the other hand if

$$\sum_{Q \in \tilde{\mathcal{Q}}, Q \subset [-\frac{1}{4\sqrt{m}}, \frac{1}{4\sqrt{m}}]^m} |Q| \geq \frac{1}{2} \frac{1}{(2\sqrt{m})^m}$$

then (16) gives

(19) $$\sum_{Q \in \mathcal{G}} |Q| \geq 8^{-m} \frac{1}{2} \frac{1}{(2\sqrt{m})^m}.$$

Now let $\mathcal{P}$ be the collection $(\widetilde{\mathcal{Q}} \setminus \mathcal{G}) \cup (\cup_{Q \in \mathcal{G}} \mathcal{P}_Q^0)$. Then $\mathcal{P}$ covers F_0 and by (15) and (16) we have

$$\sum_{Q \in \mathcal{P}} |Q| = \sum_{Q \in \widetilde{\mathcal{Q}} \setminus \mathcal{G}} |Q| + \sum_{Q \in \mathcal{G}} \sum_{P \in \mathcal{P}_Q^0} |P|$$
$$\leq \sum_{Q \in \widetilde{\mathcal{Q}} \setminus \mathcal{G}} |Q| + (1 - \theta \delta^m) \sum_{Q \in \mathcal{G}} |Q|$$
$$= \sum_{Q \in \widetilde{\mathcal{Q}}} |Q| - \theta \delta^m \sum_{Q \in \mathcal{G}} |Q|$$
$$\leq |Q_0| - \theta \delta^m \sum_{Q \in \mathcal{G}} |Q|$$

for suitable $\theta = \theta(m) \in (0,1)$, and hence by (19)

(20) $$\sum_{Q \in \mathcal{P}} |Q| \leq (1 - C^{-1} \delta^m)|Q_0|, \quad C = C(m).$$

Also note that by (4), (5), and (14) we have

(21) $$\forall P \in \mathcal{P}, \ \exists B \in \mathcal{B}_0 \text{ such that } B \subset P^{(N)} \text{ with } N \leq C\delta^{-1}$$
$$\text{and } C^{-1}\delta \text{ diam } B \leq e(P) \leq 2 \text{ diam } B,$$

where $C = C(m)$. Thus, regardless of whether or not (17) holds, we in any case get a collection of cubes covering F_0 such that (20) and (21) hold. (If (17) holds then (18) shows that we can get (20), (21) with $\mathcal{P} = \widetilde{\mathcal{Q}}$.)

Now for any $\gamma > 0$ we can trivially cover $\overline{B}_1(0) \setminus Q_0$ by a family of balls $\{B_{\sigma_k}(z_k)\}$ with $\sigma_k \geq \beta$ and $\sum_k \omega_m \sigma_k^m \leq (1+\gamma)(|B_1(0)| - |Q_0|)$, where $\beta = \beta(m, \gamma) > 0$. Similarly, there is a cover of any given cube $Q \subset \mathbb{R}^m$ by balls $\{B_{\sigma_k}(z_k)\}$ such that $\sum_k \omega_m \sigma_k^m \leq (1+\gamma)|Q|$ and $\min_k \sigma_k \geq \beta e(Q)$, with $\beta = \beta(\gamma, m) > 0$. Then (20), (21) imply that there is a cover of F by a collection $\mathcal{U}_0 = \{B_{\rho_k}(y_k)\}$ of balls with

(22) $$\sum_k \rho_k^m \leq (1-\theta)\rho_0^m, \quad \theta = \theta(\delta, m) \in (0,1),$$

and, also, for each $B_{\rho_k}(y_k) \in \mathcal{U}_0$ there is a ball $\overline{B}_{\sigma_k}(z_k) \in \mathcal{B}$ such that

(23) $$B_{\theta^{-1}\rho_k}(y_k) \supset B_{\sigma_k}(z_k).$$

To complete the proof of Lemma 2 we need the following lemma:

4.2. A general rectifiability lemma

Lemma 3 *For each $\eta \in (0, \frac{1}{16}]$, each ball $B_{\rho_0}(x_0) \subset \mathbb{R}^m$, and each non-empty subset $F \subset \overline{B}_{\rho_0}(x_0)$, there is a finite collection $\mathcal{B} = \{\overline{B}_{\rho_k}(y_k)\}_{k=1,\ldots,Q}$ of balls with centers $y_k \in F$, $F \subset \cup_{k=1}^{Q} \overline{B}_{\rho_k}(y_k)$, $\rho_k \geq \eta \rho_0 \ \forall k$, and $\sum_{k=1}^{Q} \rho_k^m \leq (1 + C\eta^\alpha)\rho_0^m$, where $C > 0$ and $\alpha \in (0, 1)$ depend only on m.*

Before we give the proof of this we show how it is used to complete the proof of Lemma 2. Let $\mathcal{U}_0$ be as in (22), (23) above, and let $B \in \mathcal{B}$ be arbitrary. According to the above lemma (with B, $F \cap B$, $C^{-1/\alpha}\theta^{1/\alpha}$ in place of B, F, η respectively), we can find a finite collection $\mathcal{R}_B$ of balls with centers in F and covering $F \cap B$ such that $\sum_{\widetilde{B} \in \mathcal{R}_B} |\widetilde{B}| \leq (1 + \theta)|B|$, and with diam $\widetilde{B} \geq \gamma$ diam B, where $\gamma = \gamma(m, \delta) > 0$. Then taking $\mathcal{U} = \cup_{B \in \mathcal{U}_0} \mathcal{R}_B$, we evidently have by (22), (23) that

$$\sum_{B \in \mathcal{U}} (\text{radius } B)^m \leq (1 + \theta)(1 - \theta)\rho_0^m \equiv (1 - \theta^2)\rho_0^m,$$

and, also, for each $\overline{B}_{\rho_k}(y_k) \in \mathcal{U}$ there is a ball $\overline{B}_{\sigma_k}(z_k) \in \mathcal{B}$ such that

$$B_{\gamma^{-1}\rho_k}(y_k) \supset B_{\sigma_k}(z_k),$$

with γ depending only on m, δ. This completes the proof of Lemma 2, subject to the lemma, the proof of which we now give:

Proof of Lemma 3: By translation and scaling it is evidently enough to check the lemma in the special case when $x_0 = 0$ and $\rho_0 = 1$. So suppose $F \subset \overline{B}_1(0)$, and let $\mathcal{Q}$ denote the set of all balls $\overline{B}_\rho(y)$ with either $y \in F$ or $F \cap \overline{B}_\rho(y) = \emptyset$. We claim that for each $k = 1, 2, \ldots$, there is a collection $\mathcal{Q}_k \subset \mathcal{Q}$ with

(1) $$\left| B_1(0) \setminus \left(\bigcup_{B \in \mathcal{Q}_k} B \right) \right| \leq (1 - \tfrac{1}{32^m})^k |B_1(0)|$$

and

(2) $$\text{dist}(B^{(1)}, B^{(2)}) \geq \tfrac{1}{16^k}, \quad B \subset \overline{B}_{1-1/16^k}(0), \quad \text{radius } B \geq \tfrac{1}{16^k}$$

for every $B, B^{(1)}, B^{(2)} \in \mathcal{Q}_k$ with $B^{(1)} \neq B^{(2)}$.

This is correct for $k = 1$ (indeed we can take $\mathcal{Q}_1$ to consist of just one ball of radius $\frac{1}{4}$), so we assume $k \geq 1$ and that (as an inductive hypothesis) $\mathcal{Q}_k \subset \mathcal{Q}$ as in (1), (2) already exists.

Let $\mathcal{P}_k$ be a maximal pairwise disjoint collection of closed balls of radius $\frac{1}{2} \cdot \frac{1}{16^k}$ with centers in $B_1(0) \setminus (\cup_{B \in \mathcal{Q}_k} B)$. Then

(3) $$B_1(0) \setminus \left(\bigcup_{B \in \mathcal{Q}_k} B \right) \subset \bigcup_{P \in \mathcal{P}_k} (\text{2-times enlargement of } P)$$

by the maximality of the collection $\mathcal{P}_k$, where the 2-times enlargement of a ball $\overline{B}_\rho(y)$ means $\overline{B}_{2\rho}(y)$.

Also, keeping in mind that no open ball of radius $\frac{1}{2} \cdot \frac{1}{16^k}$ can intersect more than one of the sets in the collection $\mathcal{Q}_k \cup \{\mathbb{R}^m \setminus B_1(0)\}$ (by (2)), we can easily check that each $P \in \mathcal{P}_k$ contains a ball P_1 of radius $\frac{1}{4} \cdot \frac{1}{16^k}$ which does not intersect the set $(\mathbb{R}^m \setminus B_1(0)) \cup (\cup_{B \in \mathcal{Q}_k} B)$. On the other hand for any ball $B_\rho(y) \subset \mathbb{R}^m$ we can find $B_{\rho/2}(\tilde{y}) \subset B_\rho(y)$ with $\overline{B}_{\rho/2}(\tilde{y}) \in Q$. (Just take $\tilde{y} = y$ in case $F \cap \overline{B}_{\rho/2}(y) = \emptyset$, and take $\tilde{y} \in F \cap \overline{B}_{\rho/2}(y)$ arbitrary in case $F \cap \overline{B}_{\rho/2}(y) \neq \emptyset$.) Then the ball P_1 contains a ball $P_2 \in Q$ with radius $\frac{1}{8} \cdot \frac{1}{16^k}$ and hence the ball $\widetilde{P}$ with the same center as P_2 and radius $\frac{1}{16^{k+1}}$ has the properties

$$\widetilde{P} \in Q, \quad \text{dist}\left(\widetilde{P}, (\mathbb{R}^m \setminus B_1(0)) \cup \left(\bigcup_{B \in \mathcal{Q}_k} B\right) \cup \left(\bigcup_{Q \in \mathcal{P}_k, Q \neq P} Q\right)\right) \geq \frac{1}{16^{k+1}}.$$

Thus we can define

$$\mathcal{Q}_{k+1} = \mathcal{Q}_k \cup \widetilde{\mathcal{P}}_k,$$

where $\widetilde{\mathcal{P}}_k = \{\widetilde{P} : P \in \mathcal{P}_k\}$, and then $\mathcal{Q}_{k+1}$ has the required properties (2) with $k+1$ in place of k.

Furthermore, for any $P \in \mathcal{P}_k$, the 32-times enlargement of $\widetilde{P}$ (i.e. $\overline{B}_{32\rho}(y)$ if $\widetilde{P} = \overline{B}_\rho(y)$) contains the 2-times enlargement of P, so, by (3),

$$B_1(0) \setminus \left(\bigcup_{B \in \mathcal{Q}_k} B\right) \subset \bigcup_{P \in \mathcal{P}_k} (\text{32-times enlargement of } \widetilde{P}),$$

and hence

$$\left| B_1(0) \setminus \left(\bigcup_{B \in \mathcal{Q}_k} B\right) \right| \leq (32)^m \sum_{P \in \mathcal{P}_k} |\widetilde{P}|.$$

Then

$$\left| B_1(0) \setminus \left(\bigcup_{B \in \mathcal{Q}_{k+1}} B\right) \right| = \left| B_1(0) \setminus \left(\bigcup_{B \in \mathcal{Q}_k} B\right) \right| - \sum_{P \in \mathcal{P}_k} |\widetilde{P}|$$

$$\leq (1 - \tfrac{1}{32^m}) \left| B_1(0) \setminus \left(\bigcup_{B \in \mathcal{Q}_k} B\right) \right|$$

$$\leq (1 - \tfrac{1}{32^m})^{k+1} |B_1(0)|$$

by the inductive hypothesis.

Thus the existence of the required collections $\mathcal{Q}_k$ as in (1), (2) is established for all k; further the above proof also establishes that for each k we can select a collection (Viz. $\widetilde{\mathcal{P}}_k \equiv \{\widetilde{P} : P \in \mathcal{P}_k\}$) of balls of radius $\frac{1}{16^{k+1}}$, each contained in $B_1(0) \setminus (\cup_{B \in \mathcal{Q}_k} B)$, such that the 32-times enlargements (each of radius $\frac{2}{16^k}$) cover all of $B_1(0) \setminus (\cup_{B \in \mathcal{Q}_k} B)$ and have total measure $\leq (32)^m |B_1(0) \setminus (\cup_{B \in \mathcal{Q}_k} B)| \leq (32)^m (1 - \tfrac{1}{32^m})^k |B_1(0)|$. Now for each $P \in \mathcal{P}_k$ such that the 32-times enlargement of $\widetilde{P}$ intersects F we can choose $\widehat{P}$ with center in F, radius 64-times the radius of $\widetilde{P}$

4.2. A general rectifiability lemma

and containing the 32-times enlargement of $\widetilde{P}$. Let the set of all such $\widehat{P}$ be denoted $\widehat{\mathcal{P}}_k$. Then $\mathcal{Q}_k \cup \widehat{\mathcal{P}}_k$ cover F and the sum of the volumes is

(4) $$\leq \left(1 + (64)^m (1 - \tfrac{1}{32^m})^k\right) |B_1(0)|.$$

We now choose k such that $16^{-k} \in [\eta, 16\eta)$. Then $(1 - \tfrac{1}{32^m})^k = ((1 - \tfrac{1}{32^m})^{qk})^{1/q} \leq (\tfrac{1}{16^k})^{1/q} \leq 16^{1/q}\eta^{1/q}$, provided q is selected such that $(1 - \tfrac{1}{32^m})^q \leq \tfrac{1}{16}$, so by (4) we see that Lemma 3 is proved with $\alpha = q^{-1}$, $C = 64^{m+1}$, and $\mathcal{B}$ equal to the subcollection consisting of all balls B in $\mathcal{Q}_k \cup \widehat{\mathcal{P}}_k$ such that $F \cap B \neq \varnothing$. □

Proof of the Rectifiability Lemma: We assume first that S has no $\tfrac{\delta}{20}$-gap in $B_{\rho_0}(x_0)$, and hence (Cf. Remark (2) above), for $\varepsilon \leq \varepsilon_0$, with $\varepsilon_0 = \varepsilon_0(n, m, \delta)$ small enough,

(1) $\quad S$ has no δ-gap in any ball $B_\tau(y)$, $y \in \overline{B}_{\rho_0/2}(x_0) \cap S$, $\tau \in [\tfrac{\rho_0}{16}, \tfrac{\rho_0}{2}]$,

and also the hypotheses I(a), (b) must hold. Let

(2) $$S^{(1)} = \{y \in S \cap \overline{B}_{\rho_0/2}(x_0) \setminus (\cup \mathcal{F}_0) : S \text{ has no } \delta\text{-gap in } B_\rho(y) \,\forall \rho \in (0, \tfrac{\rho_0}{2}]\},$$

and let

(3) $$E_1 = S \cap \overline{B}_{\rho_0/2}(x_0) \setminus (S^{(1)} \cup (\cup \mathcal{F}_0)).$$

For each $y \in S^{(1)}$ we have by the property (b) that S satisfies the uniform cone condition

(4) $\quad S \cap B_\rho(y) \subset$ the $(\varepsilon\rho)$-neighbourhood of $y + L_0$

for each $\rho \in (0, \tfrac{\rho_0}{2}]$, and hence we deduce immediately that

(5) $$S^{(1)} \subset G_1, \quad \mathcal{H}^m(S^{(1)}) \leq \omega_m \left(\tfrac{\rho_0}{2}\right)^m,$$

where G_1 is the graph of a Lipschitz function f defined over the m-dimensional subspace L_0 with $\text{Lip}\, f \leq C\varepsilon$. Let P_0 be the orthogonal projection of $\mathbb{R}^n$ onto L_0. By (1) and (i) we have that for each $y \in E_1$

(6) $\quad S$ has a $\tfrac{\delta}{2}$-gap in $B_{\rho_y}(y)$, S has no δ-gap in $B_\tau(y) \,\forall \tau \in (\rho_y, \tfrac{\rho_0}{2}]$,

where $\rho_y = \gamma(y, \tfrac{\rho_0}{2}, \delta) \in (0, \tfrac{\rho_0}{16}]$, and by hypothesis I(b)

(7) $$S \subset \{x \in B_{\rho_y}(y) : \text{dist}(x, y + L_0) \leq \varepsilon\rho_y\} \cup K_{y,\varepsilon,L_0},$$

where K_{y,ε,L_0} is the double cone $\{x \in \mathbb{R}^n : \text{dist}(x, y + L_0) \leq \varepsilon|x - y|\}$.

Next, define $F = P_0(E_1)$, and let $\mathcal{B}$ be the collection of balls in L_0 which are orthogonal projections of the balls $\overline{B}_{\rho_y}(y)$, $y \in E_1$. For the remainder of this argument, balls in L_0 will be denoted $B_\rho^0(y)$. Thus $\mathcal{B} = \{\overline{B}_{\rho_y}^0(\tilde{y})\}_{y \in E_1}$, where $\tilde{y}$ is the orthogonal projection of y onto L_0. By virtue of (6), (7) we know that F has a $\frac{\delta}{4}$-gap in each of the balls of $\mathcal{B}$, and hence by the covering lemma (Lemma 2 above) we have that there is a collection of balls $\widetilde{\mathcal{B}} = \{\overline{B}_{\rho_k}^0(\tilde{y}_k)\}$, with $\tilde{y}_k = P_0(y_k)$, $y_k \in E_1$, which cover F and which satisfy $\sum_k \rho_k^m \leq (1-\theta)\rho_0^m$, and for each k there is $\overline{B}_{\sigma_k}^0(\tilde{z}_k) = P_0(\overline{B}_{\sigma_k}(z_k)) \in \mathcal{B}$ (with $\sigma_k = \rho_{z_k}$), such that $B_{\sigma_k}^0(\tilde{z}_k) \subset B_{\theta^{-1}\rho_k}^0(\tilde{y}_k)$. By (7) with $y = z_k$ we have $S \subset \{x \in B_{\sigma_k}(z_k) : \mathrm{dist}(x, z_k + L_0) \leq \varepsilon\sigma_k\} \cup K_{z_k,\varepsilon,L_0}$, and hence in particular $|y_k - z_k| \leq 2\theta^{-1}\rho_k$, $\mathrm{dist}(y_k, z_k + L_0) \leq 2\theta^{-1}\varepsilon\rho_k$, and

$$E_1 \cap P_0^{-1}(B_{\rho_k}^0(\tilde{y}_k)) \subset E_1 \cap B_{(1+6\theta^{-1}\varepsilon)\rho_k}(y_k),$$

provided ε is sufficiently small relative to θ.

Thus, provided S has no $\frac{\delta}{20}$-gap in $B_{\rho_0}(x_0)$ and ε is sufficiently small relative to θ, we have constructed a countable collection of balls $\{B_{\tau_k}(y_k)\}$ ($\tau_k = (1+6\theta^{-1}\varepsilon)\rho_k$) with centers in E_1 such that, after a change of notation (replacing θ by $2^{m+2}\theta$)

(8)
$$E_1 \subset \bigcup_k B_{\tau_k}(y_k), \quad \sum_k \tau_k^m \leq (1 - 2^{m+2}\theta)\left(\frac{\rho_0}{2}\right)^m + C\varepsilon\rho_0^m \leq (1 - 2^{m+1}\theta)\left(\frac{\rho_0}{2}\right)^m$$

for suitable $\theta = \theta(n, m, \delta) \in (0, \frac{1}{2})$.

Now with $F = P(S \setminus B_{\rho_0/2}(x_0))$, P the orthogonal projection of $\mathbb{R}^n$ onto the affine space L_{x_0,ρ_0}, we can first cover all of $L_{x_0,\rho_0} \cap B_{\rho_0}(x_0) \setminus B_{\rho_0/2}(x_0)$ by balls $B_{\rho_k}(y_k)$ with centers y_k in L_{x_0,ρ_0} and radii $\rho_k \geq C^{-1}\rho_0$, with $C = C(m, \theta) > 0$, such that

(9)
$$\sum_k \rho_k^m \leq \left(1 + \frac{\theta}{2}\right)\left(\rho_0^m - \left(\frac{\rho_0}{2}\right)^m\right).$$

But then we can apply Lemma 3 to each of $B_{\rho_k}(y_k) \cap F$ to get a new collection $\widetilde{\mathcal{B}}_0 = \{B_{\sigma_k}(z_k)\}$ with $z_k \in F$, $F \subset \cup \widetilde{\mathcal{B}}_0$, and such that $\sum_k \sigma_k^m \leq (1+\theta)(\rho_0^m - (\rho_0/2)^m)$ and $\sigma_k \geq C^{-1}$, with $C = C(m, \theta)$. Using property (i) we can then use the $\widetilde{\mathcal{B}}_0$ to construct a collection $\mathcal{B}_1 = \{B_{(1+C\varepsilon)\sigma_k}(\hat{z}_k)\}$ with $P(\hat{z}_k) = z_k$ and $\hat{z}_k \in S \setminus B_{\rho_0/2}(x_0)$ (so $|z_k - \hat{z}_k| < \varepsilon\rho_0$ by (i)) which covers all of $S \setminus B_{\rho_0/2}(x_0)$. Thus $\widetilde{\mathcal{B}} = \{B_{\tau_k}(y_k)\} \cup \mathcal{F}_0 \cup \mathcal{B}_1$ is a collection of balls with centers in $S \cap \overline{B}_{\rho_0}(x_0)$ with the the properties that

(10)
$$S \setminus S^{(1)} \subset \bigcup_{B \in \widetilde{\mathcal{B}}} B$$

$$\sum_{B \in \widetilde{\mathcal{B}}} (\mathrm{diam}\, B)^m \leq (1 - 2^{m+1}\theta)\left(\frac{\rho_0}{2}\right)^m + C\varepsilon\rho_0^m + (1+\theta)\left(\rho_0^m - \left(\frac{\rho_0}{2}\right)^m\right)$$

$$\leq (1-\theta)\rho_0^m$$

(for ε sufficiently small depending on m, θ) and

(11)
$$S^{(1)} \subset G_1 \cap \overline{B}_{\rho_0/2}(x_0), \quad \mathcal{H}^m(S^{(1)}) \leq \omega_m \rho_0^m,$$

4.2. A general rectifiability lemma

where G_1 is the graph of a Lipschitz function f over L_0.

Of course if S does have a $\frac{\delta}{20}$-gap in the ball $B_{\rho_0}(x_0)$, then using (i) and Lemma 3 it is trivial to find a cover $\widetilde{\mathcal{B}}$ of balls such that (10) and (11) hold with $S^{(1)} = \varnothing$. Thus regardless of which of the alternative hypotheses of (I) hold, we always conclude that (10) and (11) hold.

We now proceed inductively. Assume that $J \geq 1$ and $S^{(j)} \subset S$, $\{B_{\rho_{j,k}}(x_{j,k})\}_{k=1,2,\ldots}$, $j = 1, \ldots, J$, are already constructed (with $x_{j,k} \in S$) so that $S^{(0)} = \varnothing$, $\{B_{\rho_{0,k}}(x_{0,k})\} = \{B_{\rho_0}(x_0)\}$,

$$S \setminus \bigcup_{j=1}^{J} S^{(j)} \subset \bigcup_k B_{\rho_{J,k}}(x_{J,k}),$$

$\cup_{j=1}^{J} S^{(j)}$ is contained in a countable union of Lipschitz graphs, and

$$\sum_k \rho_{j,k}^m \leq (1-\theta) \sum_k \rho_{j-1,k}^m \text{ and } \mathcal{H}^m(S^{(j)}) \leq \omega_m \sum_k \rho_{j-1,k}^m, \quad j = 1, \ldots, J.$$

Then we repeat the argument described above, starting with $S \cap \overline{B}_{\rho_{J,k}}(x_{J,k})$ in place of S and with $x_{J,k}$, $\rho_{J,k}$ in place of x_0, ρ_0. Then conclusions (10), (11) imply that we have a Lipschitz graph Σ_k^J and a subset $S_k^{(J)} \subset S \cap \Sigma_k^J \cap \overline{B}_{\rho_{k,J}/2}(x_{k,J})$ such that $\mathcal{H}^m(S_k^{(J)}) \leq \omega_m \rho_{J,k}^m$, and balls $\{B_{\rho_{J,k,\ell}}(x_{J,k,\ell})\}_{\ell=1,2,\ldots}$ with centers in S such that

$$S \cap \overline{B}_{\rho_{J,k}}(x_{J,k}) \setminus S_k^{(J)} \subset \bigcup_\ell B_{\rho_{J,k,\ell}}(x_{J,k,\ell})$$

and

$$\sum_\ell \rho_{J,k,\ell}^m \leq (1-\theta) \rho_{J,k}^m.$$

Relabelling so that $\{B_{\rho_{J,k,\ell}}(x_{J,k,\ell})\}_{k,\ell=1,2,\ldots} = \{B_{\rho_{J+1,k}}(x_{J+1,k})\}_{k=1,2,\ldots}$ and defining $S^{(J+1)} = \cup_k S_k^{(J)}$ we then have

$$\mathcal{H}^m(S^{(J+1)}) \leq \sum_k \mathcal{H}^m(S_k^{(J)}) \leq \omega_m \sum_k \rho_{J,k}^m$$

$$\sum_k \rho_{J+1,k}^m \leq (1-\theta) \sum_k \rho_{J,k}^m,$$

and

$$S \setminus \bigcup_{j=1}^{J+1} S^{(j)} \subset \bigcup_k (S \cap B_{\rho_{J,k}}(x_{J,k}) \setminus S^{(J+1)}) \subset \bigcup_k B_{\rho_{J+1,k}}(x_{J+1,k}).$$

Thus such a collection exists for all J and

$$S \setminus \bigcup_{j=1}^{J} S^{(j)} \subset \bigcup_k B_{\rho_{J,k}}(x_{J,k}), \quad \sum_k \rho_{J,k}^m \leq (1-\theta)^J \rho_0^m$$

$$\bigcup_{j=1}^{J} S^{(j)} \subset \text{ a countable union of Lipschitz graphs}$$

$$\mathcal{H}^m\left(\bigcup_{j=1}^{J} S^{(j)}\right) \leq \sum_{j=1}^{J} (1-\theta)^{j-1} \omega_m \rho_0^m.$$

Thus $S \setminus (\cup_j S^{(j)})$ has $\mathcal{H}^m$-measure zero, $\cup_j S^{(j)}$ is contained in a countable union of Lipschitz graphs, and $\mathcal{H}^m(\cup_j S^{(j)}) \leq C \rho_0^m$. Thus the lemma is proved. □

4.3 Gap Measures on Subsets of $\mathbb{R}^n$

Here we want to establish the existence of a certain class of Borel measures on closed subsets $S \subset \overline{B}_1(0)$ having the same ε-approximation property as the set S in the previous section; the main result appears in Lemma 1 below.

Let $\varepsilon > 0$ and $m \in \{1, 2, \ldots, n-1\}$, $B_\rho(z) = \{x \in \mathbb{R}^n : |x - z| < \rho\}$, and let $0 \in S \subset \overline{B}_1(0)$ be an arbitrary non-empty closed subset of $\mathbb{R}^n$ with the same ε-approximation property satisfied by S in Section 4.2. Thus for each $z \in S$ and $\rho \in (0, 1]$ there is an m-dimensional affine space $L_{z,\rho}$ containing z such that

(i) $\qquad B_\rho(z) \cap S \subset \text{ the } (\varepsilon\rho)\text{-neighbourhood of } L_{z,\rho}.$

We henceforth fix these spaces $L_{z,\rho}$, and assume $\varepsilon \leq \frac{\delta}{2} \in (0, \frac{1}{32})$.

(ii) Definition: If $z \in S$ the "δ-radius" $\rho_z \in [0, \frac{1}{4}]$ of z is defined by

$$\rho_z = \sup(\{0\} \cup \{\sigma \in (0, \tfrac{1}{4}] : \text{either } S \text{ has a } \delta\text{-gap or a } \delta\text{-tilt in } B_\sigma(z)\}),$$

where $L_{z,\sigma}$ are as in (i). Here "δ-gap" is as described in the previous section, and S is said to have a δ-tilt in $B_\sigma(z)$ if $\|(L_{z,\sigma} - z) - L_{0,1}\| \geq \delta$.

Remark: Thus ρ_z is such that S has no δ-gaps or δ-tilts in the balls $B_\tau(z)$, $\rho_z < \tau \leq \frac{1}{4}$, and, if $\rho_z > 0$, S does have a δ-gap or δ-tilt in $B_{\tau_j}(z)$ for some sequence $\tau_j \uparrow \rho_z$.

Now we are going to define a family of subsets $\{T_\rho\}_{\rho \in (0, \frac{1}{4}]}$ as follows:

(iii) Definition: For $\rho \in (0, \frac{1}{4}]$, we define open subsets $T_\rho \subset \mathbb{R}^n$ by

$$T_\rho = \bigcup_{\{z \in S : \rho_z < \rho\}} B_\rho(z),$$

4.3. Gap Measures on Subsets of $\mathbb{R}^n$

where ρ_z is as in (ii) above.

(iv) Remarks: (1) The sets T_ρ depend on S and δ, but for convenience this is not indicated by the notation. Of course $T_\rho \subset \cup_{z \in S} B_\rho(z) = S_\rho \equiv \{x \in \mathbb{R}^n : \text{dist}(x, S) < \rho\}$, so we can think of T_ρ as being some sort of refinement or reduction of S_ρ, taking into account δ-gaps and δ-tilts. Also,

$$T_\rho \subset T_{1/16} \subset B_{15/16}(0) \quad \forall \rho \in (0, \tfrac{1}{16}],$$

because S trivially has a $\frac{1}{16}$-gap in each ball $B_{1/4}(z)$ with $z \in S \setminus B_{7/8}(0)$ by virtue of the fact that $S \subset \overline{B}_1(0)$. Thus $\rho_z = \frac{1}{4}$ for $z \in S \setminus B_{7/8}(0)$, and hence, if $\rho \leq \frac{1}{16}$, Definition (iii) implies $T_\rho \subset \cup_{z \in S \cap B_{7/8}(0)} B_\rho(z) \subset B_{15/16}(0)$ as claimed.

(2) It is possible to check the following properties direct from the definition of the sets T_ρ:

(a) The σ-neighbourhood of $T_\rho \subset T_{\rho+\sigma}$ for each $\rho, \sigma > 0$ with $\rho + \sigma < \frac{1}{4}$ (so that in particular we have $\text{dist}(T_\rho, \mathbb{R}^n \setminus T_{\rho+\sigma}) \geq \sigma$).

(b) $\forall z \in S \setminus T_\rho$, $\rho \in (0, \tfrac{1}{4}]$ we have $\rho \leq \rho_z$.

(b) The $\frac{\rho}{4}$-neighbourhood of $T_\rho \setminus T_{\rho/2}$ is contained in $T_{2\rho} \setminus T_{\rho/4}$, $\rho \in (0, \tfrac{1}{8}]$.

Notice that, taking $\rho = 2^{-\ell}$ and $\sigma = 2^{-k} - 2^{-\ell}$ in (a) we have in particular that

(d) $\text{dist}(T_{2^{-\ell}}, \mathbb{R}^n \setminus T_{2^{-k}}) \geq 2^{-k-1}$ for $\ell \geq k+1$, $k \geq 2$.

Proof of (a): Take any $w \in \sigma$-neighbourhood of T_ρ. Then $w \in B_\sigma(y)$ for some $y \in T_\rho$, and by definition of T_ρ there is a $z \in S$ such that $y \in B_\rho(z)$ and $\rho > \rho_z$. But then trivially $w \in B_{\rho+\sigma}(z)$ and since $\rho + \sigma > \rho_z$ this gives $w \in T_{\rho+\sigma}$ by definition. □

Proof of (b): If $\rho > \rho_z$ then $z \in B_\rho(z) \subset T_\rho$ by (ii). □

Proof of (c): By (a), the $\frac{\rho}{4}$-neighbourhood of $T_{\rho/4}$ is contained in $T_{\rho/2}$, and hence the $\frac{\rho}{4}$-neighbourhood of $\mathbb{R}^n \setminus T_{\rho/2}$ is contained in $\mathbb{R}^n \setminus T_{\rho/4}$. Also, again by (a), the $\frac{\rho}{4}$-neighbourhood of T_ρ is contained in $T_{2\rho}$. The combination of these inclusions then gives (c) as claimed. □

Lemma 1 *There is $\delta_0 = \delta_0(m, n) \in (0, \tfrac{1}{16}]$ such that if $0 < \varepsilon \leq \tfrac{\delta}{2} \leq \tfrac{\delta_0}{2}$, if S, $\{T_\sigma\}_{\sigma \in (0, \tfrac{1}{4}]}$ are as introduced above, then there is a Borel measure μ on S with the properties $\mu(S) = 1$ and, for each $\sigma \in (0, \tfrac{1}{16}]$,*

$$C^{-1} \rho^m \leq \mu(B_\rho(z) \cap S) \leq C \rho^m, \quad \rho \in [\delta^{1/2} \sigma, \tfrac{1}{16}], \quad z \in T_\sigma \cap S,$$

where $C = C(n,m)$. The measure μ has the general form

$$\mu = C\left(\delta^{m/2}\sum_{k=2}^{\infty} 2^{-mk}\sum_{j=1}^{Q_k}[\![z_{k,j}]\!] + \mathcal{H}^m \llcorner T_0\right),$$

where $[\![z]\!]$ denotes the unit mass (Dirac mass) supported at z, $T_0 = \cap_{\rho>0} T_\rho$, C depends only on n, m, and where the $z_{k,j} \in S \cap T_{2^{-k}} \setminus T_{2^{-k-1}}$, $j = 1,\ldots,Q_k$, $k \geq 2$, with

$$S \cap T_{2^{-k}} \setminus T_{2^{-k-1}} \subset \bigcup_{\ell=\max\{k-2,2\}}^{k+1} \bigcup_{j=1}^{Q_\ell} B_{\delta^{1/2}2^{-\ell}}(z_{\ell,j}), \quad k \geq 2.$$

(v) Remarks: (1) It is important for later application that C does not depend on δ, nor indeed on S. Of course one has to keep in mind that if the set S is very badly behaved (like a Koch curve for example), then the sets T_ρ can all reduce to the empty set for sufficiently small ρ, in which case the lemma has correspondingly limited content.

(2) As part of the proof, it is shown that T_0 is contained in the graph of a Lipschitz function defined over $\{0\} \times \mathbb{R}^m$ and with Lipschitz constant $\leq C\delta$, so automatically $\mathcal{H}^m \llcorner T_0$ has total measure $\leq C$.

Proof of Lemma 1: First note that we may assume

(1) $\qquad\qquad\qquad T_{1/16} \neq \emptyset,$

otherwise there is essentially nothing to prove. Also, without loss of generality we can assume

(2) $\qquad\qquad\qquad L_{0,1} = \{0\} \times \mathbb{R}^m$

(where $L_{0,1}$ denotes the affine space $L_{z,\rho}$ of (i) in the case $z = 0$, $\rho = 1$). Notice that by Remark (iv)(2)(d) above we have

(3) $\qquad \mathrm{dist}(T_{2^{-k}} \setminus T_{2^{-k-1}}, T_{2^{-\ell}} \setminus T_{2^{-\ell-1}}) \geq 2^{-\ell-2}, \quad k \geq \ell+2, \ \ell \geq 2.$

Now we choose a maximal pairwise-disjoint collection of balls $\{B_{2^{-6}}(z_j^2)\}_{j=1,\ldots,Q_2}$ with $z_j^2 \in S \cap T_{1/4} \setminus T_{1/8}$, and proceed inductively for $k \geq 3$ to choose a maximal pairwise-disjoint collection $\{B_{2^{-k-4}}(z_j^k)\}_{j=1,\ldots,Q_k}$ with $z_j^k \in (S \cap T_{2^{-k}} \setminus T_{2^{-k-1}}) \setminus (\cup_{j=1}^{Q_{k-1}} B_{2^{-k-2}}(z_j^{k-1}))$. Now notice that by (3) this automatically implies that

(4) $\qquad \{B_{2^{-k-4}}(z_j^k)\}_{j=1,\ldots,Q_k, k\geq 2}$ is a pairwise disjoint collection.

Also, by induction and Remark (iv)(2)(c), $S \cap T_{1/4} \setminus T_{1/8} \subset \cup_{j=1}^{Q_2} B_{2^{-5}}(z_j^2)$ and

(5)
$$S \cap T_{2^{-k}} \setminus T_{2^{-k-1}} \subset \bigcup_{\ell=k-1}^{k}\bigcup_{j=1}^{Q_\ell} B_{2^{-\ell-3}}(z_j^\ell),$$
$$\bigcup_{j=1}^{Q_k} B_{2^{-k-3}}(z_j^k) \subset T_{2^{-k+1}} \setminus T_{2^{-k-2}}, \quad k \geq 3.$$

4.3. Gap Measures on Subsets of $\mathbb{R}^n$

We now define

(6) $$T_0 = \bigcap_{\ell=1}^{\infty} T_{2^{-\ell}} \left(\equiv \bigcap_{\rho \in (0, \frac{1}{4}]} T_\rho\right),$$

so that we can decompose $T_{1/4}$ as a disjoint union

(7) $$T_{1/4} = \left(\bigcup_{\ell=2}^{\infty}(T_{2^{-\ell}} \setminus T_{2^{-\ell-1}})\right) \cup T_0, \quad T_0 \subset S.$$

Notice that by Remark (iv)(2)(a) we have $\overline{T}_\rho \subset T_{\rho+\sigma} \ \forall \sigma > 0$, and hence T_0 is a closed subset of S. Also, by (4) we have

$$|z_i^k - z_j^\ell| \geq 2^{-\min\{k,\ell\}-4} \text{ for all } k, \ell \geq 2, \ i \leq Q_k, j \leq Q_\ell, \ z_i^k \neq z_j^\ell,$$

and hence (using also Remark (iv)(2)(d) to give $\text{dist}(z_j^k, T_0) \geq 2^{-k-2}$)

(8) $$(B_{2^{-k-4}}(z_j^k) \setminus \{z_j^k\}) \cap (T_0 \cup \{z_i^\ell : i = 1, \ldots, Q_\ell, \ \ell \geq 2\}) = \emptyset,$$
$$j = 1, \ldots, Q_k, \ k \geq 2.$$

Now suppose that $z = (\xi, \eta) \in S \cap T_{2^{-\ell}}$ for some $\ell \geq 2$. We then have by (2), (i) and the definition of $T_{2^{-\ell}}$ that there is $\tilde{z} \in S \cap B_{2^{-\ell}}(z)$ such that $S \cap B_\rho(\tilde{z}) \subset$ the $(2\delta\rho)$-neighbourhood of $\tilde{z} + \{0\} \times \mathbb{R}^m$ for each $\rho \in [2^{-\ell}, \frac{1}{4}]$, whence

$$S \cap B_{1/4}(\tilde{z}) \subset \{w \in B_{2^{-\ell}}(\tilde{z}) : \text{dist}(w, \tilde{z} + \{0\} \times \mathbb{R}^m) \leq 2\delta 2^{-\ell}\} \cup K_{2\delta}(\tilde{z}),$$

where, here and subsequently, we use the notation that

$$K_\delta(z) = \{(x, y) : |x - \xi| \leq \delta|(x, y) - (\xi, \eta)|\}$$

(so that $K_\delta(z)$ is a circular double cone, with vertex at $z = (\xi, \eta)$ and central axis $z + \{0\} \times \mathbb{R}^m$). Then by (2) and (i) it follows immediately that

(9) $$S \subset \{(x, y) \in B_{2^{-\ell+1}}(z) : |x - \xi| < \delta 2^{-\ell+2}\} \cup K_{4\delta}(z), \quad z \in S \cap T_{2^{-\ell}}.$$

Notice that now by (8) and (9) we have

(10) $$S \subset K_{2\delta}(z), \quad z \in T_0$$
$$T_0 \cup \{z_i^\ell : i = 1, \ldots, Q_\ell, \ \ell \geq 2\} \subset K_{32\delta}(z_j^k), \quad \forall j = 1, \ldots, Q_k, \ k \geq 2$$

(where in checking the second inclusion we used (9) with $\ell = k$ and $z = z_j^k$). But this means that $T_0 \cup \{z_j^k : k \geq 2, \ j = 1, \ldots, Q_k\}$ satisfies a uniform cone condition with respect to translates of the cone $K_{32\delta}(0)$, hence there is a Lipschitz function f on $\{0\} \times \mathbb{R}^m$ with values in $\mathbb{R}^{n-m}$, $\text{Lip } f \leq C\delta$, and

(11) $$T_0 \cup \{z_j^k : j = 1, \ldots, Q_k, \ k \geq 2\} \subset G \subset \bigcap_{z \in T_0 \cup \{z_j^k : j \leq Q_k, k \geq 2\}} K_{32\delta}(z),$$

where $G = \operatorname{graph} f$. Notice that if $z \in T_{2^{-\ell}} \cap S$ then either $z \in T_0$ or $z \in S \cap T_{2^{-k}} \setminus T_{2^{-k-1}}$ for some $k \geq \ell$. In the latter case (5) implies $|z - z_j^q| < 2^{-k-2}$ for some $j \leq Q_q$, $q \geq 2$ and $q = k - 1$ or k. So in either case we have a $\tilde{z} \in T_0 \cup \{z_j^k : j \leq Q_k, \ k \geq \max(2, \ell - 1)\}$ with $|\tilde{z} - z| < 2^{-\ell}$ and hence by (9) with $\tilde{z}$ in place of z we have

$$\operatorname{dist}(z, \tilde{z} + \{0\} \times \mathbb{R}^m) \leq C\delta 2^{-\ell},$$

whence (since $\operatorname{Lip} f \leq C\delta$ and $\tilde{z} \in G$) we can conclude

(12) $$\operatorname{dist}(z, G) \leq C\delta 2^{-\ell}, \quad \forall z \in T_{2^{-\ell}} \cap S, \ \ell \geq 2.$$

Next we claim that

(13) $$T_{1/16} \cap G \setminus T_0 \subset \bigcup_{k=2}^{\infty} \bigcup_{j=1}^{Q_k} B_{2^{-k+2}}(z_j^k).$$

To see this we note that if $w \in G \cap T_{2^{-\ell}} \setminus T_{2^{-\ell-1}}$ with $\ell \geq 4$, then (by definition of $T_{2^{-\ell}}$) there is $z \in S \cap T_{2^{-\ell}}$ with $|w - z| < 2^{-\ell}$. Then (12) holds, and using (9) and the fact that S has no δ-gaps in $B_{2^{-\ell+1}}(z)$, we see that there is $z_1 \in S$ with $|z_1 - w| < C\delta 2^{-\ell} \leq 2^{-\ell-2}$. By virtue of Remark (iv)(2)(c) (with $\rho = 2^{-\ell-1}$) and the fact that $w \in T_{2^{-\ell}} \setminus T_{2^{-\ell-1}}$, we conclude that then $z_1 \in T_{2^{-\ell+1}} \setminus T_{2^{-\ell-2}}$, so by (5) we have $z_1 \in \bigcup_{|q-\ell| \leq 2} \bigcup_{j=1}^{Q_q} B_{2^{-\ell-1}}(z_j^q)$, and hence $w \in \bigcup_{|q-\ell| \leq 2} \bigcup_{j=1}^{Q_q} B_{2^{-\ell}}(z_j^q)$. Since $w \in G \cap T_{1/16} \setminus T_0$ was arbitrary, this proves (13) as claimed.

Now we define a Borel measure μ_0 on S by setting

(14) $$\mu_0 = \sum_{\ell=2}^{\infty} 2^{-m\ell} \sum_{j=1}^{Q_\ell} [\![z_j^\ell]\!] + \mathcal{H}^m \llcorner T_0,$$

where $[\![z]\!]$ denotes the unit mass (Dirac mass) supported at z. Notice that since T_0 is contained in the Lipschitz graph G, we have, for any $w \in \mathbb{R}^n$ and any $\rho \in (0, 1)$,

(15) $$(\mathcal{H}^m \llcorner T_0)(B_\rho(w)) \equiv \mathcal{H}^m(T_0 \cap B_\rho(w)) \leq C\rho^m.$$

Now assume $z \in S \cap T_{2^{-\ell}}$, $\ell \geq 2$, and $\rho \in (0, \tfrac{1}{4}]$. By definition (14) we have

(16) $$\mu_0(B_\rho(z) \setminus T_0) = \sum_{k=2}^{\infty} 2^{-mk} \sum_{z_j^k \in B_\rho(z)} 1.$$

By (3) the sums in (16) are empty for $k < \ell - 4$, and $2^{-mk} \leq C\mathcal{H}^m(B_\rho(z) \cap B_{2^{-k-2}}(z_j^k) \cap G)$ if $\rho \geq 2^{-\ell}$, $k \geq \ell - 4$ and $z_j^k \in B_\rho(z)$, so (16) and (5) imply

(17)
$$\mu_0(B_\rho(z) \setminus T_0) = \sum_{k=2}^{\infty} 2^{-mk} \sum_{z_j^k \in B_\rho(z)} 1 \leq C \sum_{k=2}^{\infty} \sum_{z_j^k \in B_\rho(z)} \mathcal{H}^m(B_\rho(z) \cap B_{2^{-k-2}}(z_j^k) \cap G)$$
$$\leq C \sum_{k=3}^{\infty} \mathcal{H}^m(B_\rho(z) \cap (T_{2^{-k+1}} \setminus T_{2^{-k-2}}) \cap G) + C\mathcal{H}^m(B_\rho(z) \cap G)$$
$$\leq C\mathcal{H}^m(B_\rho(z) \cap G).$$

4.3. Gap Measures on Subsets of $\mathbb{R}^n$

Evidently (17) and the fact that $T_0 \subset G$ then give

(18) $$\mu_0(B_\rho(z)) \leq C\mathcal{H}^m(B_\rho(z) \cap G) \leq C\rho^m, \quad z \in S \cap T_{2^{-\ell}}, \ \rho \geq 2^{-\ell}, \ \ell \geq 2.$$

On the other hand, using (16) again, since $\mathcal{H}^m(B_{2^{-k+2}}(z_j^k) \cap G \cap B_\rho(z)) \leq C2^{-mk}$ we similarly obtain for $z \in T_{2^{-\ell}} \cap S$ and $\rho \geq 2^{-\ell}$

$$\sum_{k=2}^{\infty} \sum_{z_j^k \in B_\rho(z)} \mathcal{H}^m(B_{2^{-k+2}}(z_j^k) \cap B_\rho(z) \cap G) \leq C\mu_0(B_\rho(z) \setminus T_0),$$

(where the sums on the left are taken to be zero when there are no z_j^k in $B_\rho(z)$). Then since $\mu_0 \llcorner T_0 = \mathcal{H}^m \llcorner T_0$, we obtain by summation that

$$\sum_{k=2}^{\infty} \sum_{z_j^k \in B_\rho(z)} \mathcal{H}^m(B_{2^{-k+2}}(z_j^k) \cap B_\rho(z) \cap G) + \mathcal{H}^m(B_\rho(z) \cap T_0) \leq C\mu_0(B_\rho(z)).$$

Now by virtue of (13) we have $T_{1/16} \cap G \setminus T_0 \subset \bigcup_{k=2}^{\infty} \bigcup_{j=1}^{Q_k} B_{2^{-k+2}}(z_j^k)$, and hence this last inequality gives, for any $z \in S \cap T_{2^{-\ell}}$, $\ell \geq 4$, and $\rho \in [2^{-\ell}, \frac{1}{4}]$,

(19) $$C^{-1}\rho^m \leq \mathcal{H}^m(B_\rho(z) \cap T_{1/16} \cap G)$$
$$\leq C\mu_0(B_\rho(z)),$$

where we used (12) to justify the first inequality. (Using (12) and the definition of $T_{1/16}$ it can be checked that $B_\rho(z) \cap T_{1/16}$ contains a ball of radius $\frac{\rho}{32}$ centered on G.)

Now we construct the measure μ. For each $k \geq 2$ and $j = 1, \ldots, Q_k$ we select a maximal pairwise disjoint collection of balls $B_{\delta^{1/2}2^{-k-2}}(w_i^{k,j})$, $i = 1, \ldots, R_{j,k}$, subject to the restriction that $w_i^{k,j} \in T_{1/4} \cap S \cap B_{2^{-k-3}}(z_k^j)$ (so $B_{2^{-k-3}}(w_i^{k,j}) \cap S \subset S \cap B_{2^{-k-2}}(z_k^j) \subset S \cap T_{2^{-k+1}} \setminus T_{2^{-k-2}}$ for $k \geq 3$ by Remark (iv)(2)(c)). By the ε-approximation property (i) (with $\varepsilon \leq \frac{\delta}{2}$), and by the no δ-gaps hypothesis in the definition of $T_{2^{-\ell}}$, we know that, if $\delta = \delta(m,n)$ is sufficiently small, $\bigcup_{i=1}^{R_{j,k}} B_{\delta^{1/2}2^{-k-1}}(w_i^{k,j}) \supset S \cap B_{2^{-k-3}}(z_j^k)$, and for each $\sigma \in [\delta^{1/2}, 1]$ and each $w \in S \cap B_{2^{-k-3}}(z_j^k)$ we have

(20) $$C^{-1}\left(\frac{\sigma}{\delta^{1/2}}\right)^m \leq \#\{i : w_i^{k,j} \in B_{2^{-k}\sigma}(w)\} \leq C\left(\frac{\sigma}{\delta^{1/2}}\right)^m, \quad C = C(m),$$

where $\#A$ denotes the number of points in the set A. Notice also that by (5) we have

(21) $$S \cap T_{2^{-k}} \setminus T_{2^{-k-1}} \subset \bigcup_{\ell=\max\{k-1,2\}}^{k} \bigcup_{j=1}^{Q_\ell} \bigcup_{i=1}^{R_{\ell,j}} B_{\delta^{1/2}2^{-\ell-1}}(w_i^{\ell,j}), \quad k \geq 2.$$

Then we define

$$\mu = \sum_{k \geq 2} 2^{-mk} \delta^{m/2} \sum_{j=1}^{Q_k} \sum_{i=1}^{R_{k,j}} [w_i^{k,j}] + \mathcal{H}^m \llcorner T_0,$$

and in place of (16) we have

(16') $$\sum_{k=2}^{\infty} \sum_{w_i^{k,j} \in B_\rho(z)} 2^{-mk} \delta^{m/2} = \mu(B_\rho(z) \setminus T_0),$$

for any $z \in T_{2^{-\ell}}$, $\ell \geq 2$, and any $\rho \in (0, \frac{1}{4}]$. Then by virtue of (20) and (12), we can use a minor modification of the argument leading from (16) to (18), (19) in order to deduce that

$$C^{-1} \rho^m \leq \mu(B_\rho(z)) \leq C \rho^m, \quad z \in T_{2^{-\ell}}, \; \ell \geq 4, \; \rho \in [\delta^{1/2} 2^{-\ell}, \tfrac{1}{16}].$$

By (1) we have in particular that $\mu(S) \geq C^{-1}$. Then, appropriately relabeling the $w_i^{k,j}$ and using (21), we see that μ satisfies all the conditions stated in the lemma, except that in place of the condition $\mu(S) = 1$ we have

$$C^{-1} \leq \mu(S) \leq C, \quad C = C(m, n),$$

and hence the required measure is obtained by taking a suitable multiple of μ. □

4.4 Energy Estimates

Here we continue to assume that $u \in W^{1,2}(\Omega; N)$ is an energy minimizing map, $\Omega \subset \mathbb{R}^n$ open, and throughout this section we assume that $B_\rho(0) \subset \Omega$.

Here we are often going to use the variables $(r, y) = (|x|, y)$ corresponding to a given point $(x, y) \in \mathbb{R}^3 \times \mathbb{R}^{n-3}$, and it will be convenient to introduce the additional notation

$$B_\rho^+ = \{(r, y) : r > 0, \; r^2 + |y|^2 < \rho^2\}, \quad B_\rho^+(y_0) = \{(r, y) : r > 0, \; |y - y_0|^2 < \rho^2\}$$

for given $y_0 \in \mathbb{R}^{n-3}$ and $\rho > 0$.

By a "homogeneous cylindrical map" φ (abbreviated HCM) we mean a homogeneous degree zero energy minimizing map $\varphi \in W^{1,2}_{\text{loc}}(\mathbb{R}^n; N)$ such that $\widetilde{\varphi}(x, y) \equiv \varphi_0(x)$, with $\widetilde{\varphi} = \varphi \circ q$ for some orthogonal transformation q of $\mathbb{R}^n$ and $\varphi_0 \in W^{1,2}_{\text{loc}}(\mathbb{R}^3; N)$. Of course every such φ_0 is homogeneous of degree zero on $\mathbb{R}^3$ and, since $\dim \operatorname{sing} \varphi \leq n - 3$ (by Corollary 1 of Section 3.4), we have automatically that $\varphi_0 | S^2$ is a smooth harmonic map of S^2 into N. Also, by our previous discussion (see Section 3.10) we then know that $\Delta_{\mathbb{R}^n} \widetilde{\varphi} + \sum_{j=1}^n A_{\widetilde{\varphi}}(D_j \widetilde{\varphi}, D_j \widetilde{\varphi}) \equiv (\Delta_{\mathbb{R}^n} \widetilde{\varphi})^T = 0$, and hence, since $\Delta_{\mathbb{R}^n} \widetilde{\varphi}(x, y) = r^{-2} \Delta_{S^2} \varphi_0(\omega)$ with $\omega = |x|^{-1} x$, we have

(i) $$(\Delta_{S^2} \varphi_0)^T = 0 \text{ on } S^2,$$

4.4. Energy Estimates

where, at each $w \in S^2$, $(\cdot)^T$ means orthogonal projection into $T_{\varphi_0(w)}N$.

Notice that we do not assume that φ is necessarily a tangent map of u (although the reader should keep in mind that such HCM's exist at each point of $\text{sing}_* u$ as described in Section 3.5).

The main inequality of this section is given in the following theorem:

Theorem 1 *If N is real-analytic, $\zeta \in (0, \frac{1}{4})$, and $\beta > 0$ then there are $C = C(\beta, N, n) > 0$, $\eta = \eta(\beta, N, n, \zeta) > 0$ and $\alpha = \alpha(\beta, N, n) \in (0, 1)$ such that the following holds. If $\Theta_u(0) \leq \beta$, $\rho^{2-n} \int_{B_\rho(0)} |Du|^2 - \Theta_u(0) \leq \eta$, and $\rho^{-n} \int_{B_\rho(0)} r^2(u_r^2 + u_y^2) < \eta^2$, then there is an HCM $\varphi(x, y) \equiv \varphi(x, 0)$ with $\rho^{-n} \int_{B_\rho(0)} |u - \varphi|^2 < \zeta$, $\Theta_u(0) - \zeta \leq \Theta_\varphi(0) \leq \Theta_u(0) + \zeta$, and*

$$\int_{B_{\rho/2}^+} \left| \int_{S^2} r^2(|Du|^2 - |D\varphi|^2) \, d\omega \right| r^2 dr dy \leq C \int_{B_\rho(0)} r^2(u_r^2 + u_y^2) +$$

$$+ C\rho^3(\rho^{n-3})^{1-1/(2-\alpha)} \left(\rho^{-3} \int_{B_\rho(0) \setminus \{(x,y) : |x| < \rho/2\}} r^2(u_r^2 + u_y^2) \right)^{1/(2-\alpha)}.$$

If N is merely smooth rather than real-analytic, then the same is true with $\alpha = 1$ provided that the integrability condition (xiii) of Section 3.14 holds.

In proving Theorem 1 we shall need three lemmas, each of which is of some independent interest. The first of these gives some important general facts about HCM's; we use the notation

(ii)
$$\mathcal{C}_\beta = \{\varphi_0 \in C^3(S^2; N) : \varphi(x, y) \equiv \varphi_0(|x|^{-1}x) \text{ is an HCM and } \mathcal{E}_{S^2}(\varphi_0) \leq \beta\}$$

Then we have the following:

Lemma 1 *Suppose N is real-analytic. For each $\beta > 0$, $\mathcal{C}_\beta$ is compact in $C^3(S^2; N)$, and there is $\zeta_1 = \zeta_1(\beta, N, n) \in (0, \frac{1}{4}]$ such that if $\varphi_1, \varphi_2 \in \mathcal{C}_\beta$ and $\|\varphi_1 - \varphi_2\|_{L^2} < \zeta_1$ then $\mathcal{E}_{S^2}(\varphi_1) = \mathcal{E}_{S^2}(\varphi_2)$. Furthermore there are constants $\zeta_2 = \zeta_2(\beta, N, n) > 0$ and $\alpha = \alpha(\beta, N, n) \in (0, 1)$ such that*

$$|\mathcal{E}_{S^2}(\psi) - \mathcal{E}_{S^2}(\varphi)|^{2-\alpha} \leq C \int_{S^2} |(\Delta_{S^2}\psi)^T|^2, \quad C = C(\beta, N, n),$$

whenever $\varphi \in \mathcal{C}_\beta$ and $\psi \in C^3(S^2; N)$ with $|\psi - \varphi|_{C^3} < \zeta_2$. If N is merely smooth rather than real-analytic, then the same continues to hold, with best exponent $\alpha = 1$, in case all the $\varphi \in \mathcal{C}_\beta$ satisfy the integrability condition (xiii) of Section 3.14.

Remark: Thus we have uniform Łojasiewicz inequality for a whole C^3 neighbourhood of $\mathcal{C}_\beta$, and also, by the first part of the above lemma, $\mathcal{E}_{S^2}(\varphi)$ is constant on the connected components of the set $\mathcal{C}_\beta$ and there are only finitely many values of $\mathcal{E}_{S^2}(\varphi)$ corresponding to $\varphi \in \mathcal{C}_\beta$.

Proof of Lemma 1: The compactness of $\mathcal{C}_\beta$ is a direct consequence of the estimates of Section 3.6 and the compactness theorem of Section 3.6. Next suppose there is no such ζ_1. Then there must be sequences φ_j, $\widetilde{\varphi}_j$ in $\mathcal{C}_\beta$ converging in $C^3(S^2; N)$ to a common limit φ but with

(1) $$\mathcal{E}_{S^2}(\varphi_j) \neq \mathcal{E}_{S^2}(\widetilde{\varphi}_j) \ \forall j.$$

But then according to the Lojasiewicz inequality of Section 3.14 we have $\alpha = \alpha(\varphi) \in (0,1)$ and $\sigma = \sigma(\varphi) > 0$ such that

(2) $$|\mathcal{E}_{S^2}(\psi) - \mathcal{E}_{S^2}(\varphi)|^{2-\alpha} \leq C\|(\Delta\psi)^T\|^2_{L^2(S^2)}, \quad \psi \in C^3(S^2; N),\ |\psi - \varphi|_{C^3} < \sigma,$$

where $C = C(n, N, \varphi)$. But then for all sufficiently large j we can apply this with $\psi = \varphi_j$, $\widetilde{\varphi}_j$ in order to deduce that $\mathcal{E}_{S^2}(\varphi_j) = \mathcal{E}_{S^2}(\widetilde{\varphi}_j) = \mathcal{E}_{S^2}(\varphi)$, thus contradicting (1).

Now if the inequality of the lemma fails, then there are sequences $\varphi_j \in \mathcal{C}_\beta$ and $\psi_j \in C^3(S^2; N)$ both converging to a given $\varphi \in \mathcal{C}_\beta$ but such that

(3) $$|\mathcal{E}_{S^2}(\varphi_j) - \mathcal{E}_{S^2}(\psi_j)|^{1-\alpha_j/2} \leq C\|(\Delta\psi_j)^T\|_{L^2},$$

where $\alpha_j \downarrow 0$ as $j \to \infty$. But then $\mathcal{E}_{S^2}(\varphi_j) = \mathcal{E}_{S^2}(\varphi)$ for all sufficiently large j by the first part of the proof above, and (3) contradicts (2). $\square$

Lemma 2 *If $B_\sigma(0) \subset \Omega$, if $\beta > 0$ and if $\sup_{B_{7\sigma/8}(0) \setminus \{(x,y) : |x| \leq \sigma/16\}} \sum_{j=0}^3 \sigma^j |D^j u| \leq \beta$, then*

$$\sup_{B^+_{3\sigma/4} \setminus \{(r,y) : r \leq \sigma/8\}} \int_{S^2} |(\Delta_{S^2} u(r,y))^T|^2\, d\omega \leq C\sigma^{-n} \int_{B_\sigma \setminus \{(x,y) : |x| \leq \sigma/16\}} r^2(u_r^2 + u_y^2)$$

and

$$\sup_{B^+_{3\sigma/4} \setminus \{(r,y) : r \leq \sigma/8\}} \left| r\nabla_{r,y} \int_{S^2} |\nabla^{S^2} u(r,y)|^2\, d\omega \right| \leq C\sigma^{-n} \int_{B_\sigma \setminus \{(x,y) : |x| \leq \sigma/16\}} r^2(u_r^2 + u_y^2).$$

Here $\nabla_{r,y}$ means the gradient with respect to the variables $(r,y) \in B^+_\sigma$, $C = C(\beta, N, n)$, and $u(r,y)$ denotes the function on S^2 defined by $u(r,y)(\omega) = u(r\omega, y)$.

Proof: As discussed in Section 3.10, the Euler-Lagrange system for u says exactly that $(\Delta_{\mathbb{R}^n} u)^T = 0$ (meaning that $\tau \cdot \Delta_{\mathbb{R}^n} u = 0$ for any $\tau \in T_u N$). Since

$$(\Delta_{\mathbb{R}^n} u)(r,y) = r^{-2} \Delta_{S^2} u(r,y) + r^{-2} \frac{\partial}{\partial r}\left(r^2 \frac{\partial u(r,y)}{\partial r}\right) + \sum_{j=1}^{n-3} \frac{\partial^2 u(r,y)}{\partial y^{j\,2}},$$

4.4. Energy Estimates

we thus have

(1) $$r^{-2}(\Delta_{S^2}u)^T = -(\Delta_{r,y}u)^T,$$

where $\Delta_{r,y}u = r^{-2}\frac{\partial}{\partial r}(r^2\frac{\partial u(r,y)}{\partial r}) + \sum_{j=1}^{n-3}\frac{\partial^2 u(r,y)}{\partial y^{j\,2}}$.

Next we note that by differentiating the equation $\Delta_{\mathbb{R}^n}u + \sum_{j=1}^{n}A_u(D_ju, D_ju) = 0$, we obtain for $v = \partial u/\partial y^j$ a linear equation of the form

$$\mathcal{L}_u v = 0 \text{ on } B_{7\sigma/8}(0) \setminus \{(x,y) : |x| < \tfrac{\sigma}{16}\}.$$

Here (Cf. Section 3.14) $\mathcal{L}_u$ is the linear elliptic operator defined by

$$\mathcal{L}_u w \equiv \frac{d}{ds}\mathcal{M}(\Pi(u+sw))|_{s=0},$$

where $\mathcal{M}(u) = (\Delta_{\mathbb{R}^n}u)^T \equiv \Delta_{\mathbb{R}^n}u + \sum_{j=1}^{n}A_u(D_ju, D_ju)$ and where Π denotes the nearest point projection into N as in Appendix 2.12.3. Notice that, if $w(x,y) \in T_{u(x,y)}N$ on $B_{7\sigma/8}(0) \setminus \{(x,y) : |x| < \tfrac{\sigma}{16}\}$ then the operator $\mathcal{L}_u w$ has the form

$$\Delta_{\mathbb{R}^n} w + r^{-1}a \cdot \nabla w + r^{-2}b \cdot w$$

on $B_{7\sigma/8}(0) \setminus \{(x,y) : |x| < \tfrac{\sigma}{16}\}$, with $|a|, |b| \leq C(n, N, \beta)$. Then the standard $C^{1,\alpha}$ Schauder theory for such linear operators (see Section 1.7) gives

(2) $$\sup_{B_{7\sigma/9}\setminus\{(x,y):|x|<\sigma/9\}} \sum_{j=0}^{1} |\sigma^j D^j u_y|^2 \leq C\sigma^{-n}\int_{B_{8\sigma/9}\setminus\{(x,y):|x|<\sigma/10\}} u_y^2.$$

Similarly the quantity $v = ru_r + y \cdot u_y$ (i.e. $v = Ru_R$, where $R = \sqrt{r^2 + |y|^2}$) satisfies the same equation $\mathcal{L}_u v = 0$, and hence $w = ru_r$ satisfies the equation

$$\mathcal{L}_u(ru_r) = -\sum_{j=1}^{n-3} y^j \mathcal{L}_u(u_{y^j}) - 2(\Delta_y u + \sum_{j=1}^{n-3} A_u(u_{y^j}, u_{y^j})) = -2(\Delta_y u + \sum_{j=1}^{n-3} A_u(u_{y^j}, u_{y^j})).$$

Thus, again by Schauder theory (this time for the solutions of the inhomogeneous equation $\mathcal{L}_u w = f$, with $f = -2(\Delta_y u + \sum_{j=1}^{n-3} A_u(u_{y^j}, u_{y^j}))$, and using (2) to estimate $\sup |f|$, we deduce that u_r satisfies

(3) $$\sup_{B_{3\sigma/4}\setminus\{(x,y):|x|<\sigma/8\}} \sum_{j=0}^{1} |\sigma^j D^j(ru_r)|^2 \leq C\sigma^{-n}\int_{B_{3\sigma/4}\setminus\{(x,y):|x|<\sigma/16\}} r^2(u_r^2 + u_y^2).$$

Now (1), (2), (3) evidently imply the first inequality of the lemma.

Next note that by directly differentiating and integrating by parts, we have

(4) $$\frac{\partial}{\partial y^j}\int_{S^2} |\nabla^{S^2} u(r,y)|^2\, d\omega = 2\int_{S^2} \nabla^{S^2} u(r,y) \cdot \nabla^{S^2}\frac{\partial u(r,y)}{\partial y^j}$$
$$= -2\int_{S^2} \Delta_{S^2} u(r,y) \cdot \frac{\partial u(r,y)}{\partial y^j},$$

and

$$(5) \quad \frac{\partial}{\partial r} \int_{S^2} |\nabla^{S^2} u(r,y)|^2 \, d\omega = 2 \int_{S^2} \nabla^{S^2} u(r,y) \cdot \nabla^{S^2} \frac{\partial u(r,y)}{\partial r}$$

$$= -2 \int_{S^2} \Delta_{S^2} u(r,y) \cdot \frac{\partial u(r,y)}{\partial r},$$

where $u(r,y)$ is the function on S^2 defined by $u(r,y)(\omega) = u(r\omega, y)$.

So since $\partial u/\partial y^j$ and $\partial u/\partial r$ are in $T_u N$ we have from (1), (4) and (5) that

$$(6) \quad \frac{\partial}{\partial y^j} \int_{S^2} |\nabla^{S^2} u(r,y)|^2 \, d\omega = -2r^2 \int_{S^2} \Delta_{r,y} u(r,y) \cdot \frac{\partial u(r,y)}{\partial y^j},$$

and

$$(7) \quad \frac{\partial}{\partial r} \int_{S^2} |\nabla^{S^2} u(r,y)|^2 \, d\omega = -2r^2 \int_{S^2} \Delta_{r,y} u(r,y) \cdot \frac{\partial u(r,y)}{\partial r}.$$

The proof is now completed by using (2), (3) to estimate the right side. $\square$

Next we have a lemma which gives important information about approximation of u by HCM's.

Lemma 3 *If $\beta > 0$, $\rho^{2-n} \int_{B_\rho(0)} |Du|^2 \leq \beta$, then for each $\zeta > 0$ there are constants $\eta = \eta(\beta, \zeta, N, n) > 0$, $\alpha = \alpha(N, n, \beta) \in (0, 1)$ such that $\rho^{-n} \int_{B_{3\rho/4}(0) \setminus \{|x| < \rho/2\}} r^2(u_r^2 + u_y^2) < \eta$ and $\rho^{2-n} \int_{B_\rho(0)} |Du|^2 - \Theta_u(0) \leq \eta$ imply that there is an HCM φ with $\varphi(x,y) \equiv \varphi(x,0)$, $\Theta_u(0) - \zeta \leq \Theta_\varphi(0) \leq \Theta_u(0) + \zeta$, and*

$$\rho^{-n} \int_{B_{15\rho/16}(0)} |u - \varphi|^2 \leq \zeta, \quad \sup_{B_{7\rho/8}(0) \setminus \{(x,y): |x| \leq \rho/16\}} \sum_{j=0}^{3} \rho^j |D^j u - D^j \varphi|_{C^3} \leq \zeta.$$

If in addition N is real analytic, then there is $\zeta_0 = \zeta_0(n, N, \beta) \leq \min\{\zeta_1, \zeta_2\}$, ζ_1, ζ_2 as in Lemma 1, such that if $\zeta \leq \zeta_0$, then the φ from the first part above satisfies

$$\sup_{B_{3\rho/4}^+ \setminus \{(r,y): r < \rho/4\}} \left| \int_{S^2} (|\nabla^{S^2} u(r,y)|^2 - |\nabla^{S^2} \varphi|^2) \, d\omega \right| \leq$$

$$\leq C \left(\rho^{-n} \int_{B_\rho \setminus \{(x,y): |x| < \rho/8\}} r^2 (u_r^2 + u_y^2) \right)^{1/(2-\alpha)}.$$

Here, as in Lemma 2, $u(r,y)$ denotes the function on S^2 defined by $u(r,y)(\omega) = u(r\omega, y)$. If in place of the real-analyticity assumption on N we assume the integrability condition (xiii) of Section 3.14, then these additional conclusions hold with best exponent $\alpha = 1$.

4.4. Energy Estimates

Proof: First notice that by using the interpolation $\rho^j|D^j(u-\varphi)|_{C^0} \leq \varepsilon\rho^4|D^4(u-\varphi)|_{C^0} + C\varepsilon^{-4-n}\rho^{-n/2}\|u-\varphi\|_{L^2}$ (with $\varepsilon = \zeta^\gamma$) together with the regularity theorem of Section 2.3 and the estimates of Section 3.6, an inequality of the form

(1) $$\sup_{B_{7\rho/8}(0) \setminus \{(x,y) : |x| \leq \rho/16\}} \sum_{j=0}^{3} \rho^j |D^j u - D^j \varphi| \leq C\zeta^\gamma, \quad \gamma = \gamma(n) > 0,$$

is implied by the inequalities $\rho^{-n} \int_{B_{15\rho/16}(0)} |u-\varphi|^2 < \zeta$ and $\Theta_\varphi(0) \leq \Theta_u(0) + \zeta \leq \rho^{2-n} \int_{B_\rho(0)} |Du|^2 + \zeta \leq \beta + \zeta$, provided $\zeta = \zeta(n, N, \beta)$ is small enough. Hence (1) will be established if we only check the other inequalities

(2) $$\Theta_u(0) - \zeta \leq \Theta_\varphi(0) \leq \Theta_u(0) + \zeta$$
$$\rho^{-n} \int_{B_{15\rho/16}(0)} |u-\varphi|^2 < \zeta.$$

By rescaling it is enough to check (2) in case $\rho = 1$. Then if there is no such η, there must exist a sequence $u^{(j)} \in W^{1,2}(B_1(0); N)$ of energy minimizers with

(3) $$\mathcal{E}_{B_1(0)}(u^{(j)}) - \Theta_{u^{(j)}}(0) \to 0$$

and

(4) $$\int_{B_{3/4}(0) \setminus \{(x,y) : |x| \leq \frac{1}{2}\}} ((u_r^{(j)})^2 + (u_y^{(j)})^2) \to 0,$$

yet such that, for every HCM φ with $\varphi(x,y) \equiv \varphi(x,0)$, at least one of the inequalities in (2) fails with $\rho = 1$ and $u = u^{(j)}$. Notice that (3) together with monotonicity implies

(5) $$\Theta_{u^{(j)}}(0) \leq \rho^{2-n} \int_{B_\rho(0)} |Du^{(j)}|^2 \leq \int_{B_1(0)} |Du^{(j)}|^2 \leq \Theta_{u^{(j)}}(0) + \varepsilon_j, \quad \forall \rho \in (0,1),$$

where $\varepsilon_j \downarrow 0$. By (4), (5) and the compactness theorem for energy minimizers (see Section 2.9) we know that there is a subsequence (still denoted $u^{(j)}$) such that $u^{(j)} \to \varphi$ locally in $W^{1,2}(B_1; \mathbb{R}^p)$, where $\varphi_r \equiv 0$ and $\varphi_y \equiv 0$ on $B_{5/8}(0) \setminus \{(x,y) : |x| \leq \frac{1}{2}\}$ and $\rho^{2-n} \int_{B_\rho(0)} |D\varphi|^2 \equiv \lim \Theta_{u^{(j)}}(0)$ for every $\rho \in (0,1)$. But then in particular $\Theta_\varphi(0) \equiv \rho^{2-n} \int_{B_\rho(0)} |D\varphi|^2 \, \forall \rho \in (0,1)$, so by monotonicity for φ we have that φ extends to a homogeneous degree zero map in $W^{1,2}_{\text{loc}}(\mathbb{R}^n; N)$. Then since $\varphi_r \equiv 0$ and $\varphi_y \equiv 0$ on $B_{5/8}(0) \setminus \{(x,y) : |x| \leq \frac{1}{2}\}$ we have that φ is an HCM with $\varphi(x,y) \equiv \varphi(x,0)$. In view of (5) we see that in fact (2) now holds with $u = u^{(j)}$ (j sufficiently large) and $\rho = 1$, a contradiction. Thus the required inequalities (2) (and hence also (1)) must hold for some HCM φ with $\varphi(x,y) \equiv \varphi(x,0)$, provided η is sufficiently small.

We now need to establish the final inequality of the lemma. By virtue of Lemma 1 we have that there is $\alpha = \alpha(N, n, \beta) \in (0,1)$ and $\zeta = \zeta(N, n, \beta) > 0$ such that (1) implies

$$|\mathcal{E}_{S^2}(u(r,y)) - \mathcal{E}_{S^2}(\varphi)|^{2-\alpha} \leq C\|(\Delta_{S^2} u(r,y))\|^2_{L^2(S^2)}$$

for each $(r,y) \in B^+_{7\rho/8}(0)$ with $r \geq \frac{\rho}{16}$, where $C = C(n, N, \beta)$. Then the required inequality holds by virtue of Lemma 2; notice that the hypothesis

$$\sup_{B_{7\rho/8}(0) \setminus \{(x,y) : |x| \leq \rho/16\}} \sum_{j=0}^{3} \rho^j |D^j u| \leq \beta$$

required in Lemma 2 is satisfied (with $C\beta$ in place of β) by virtue of Section 3.6 and the inequality (1) above. □

We shall need the following corollary of the above lemma later. (It is not needed for the present section.)

Corollary 1 *For any given $\zeta, \beta > 0$ there is $\eta_0 = \eta_0(\zeta, \beta, n, N) > 0$ such that the following holds. Suppose φ is any HCM with $\varphi(x,y) \equiv \varphi(x,0)$ and $\Theta_\varphi(0) \leq \beta$, suppose $u \in W^{1,2}(B_\rho(0); N)$ is energy minimizing with $\rho^{2-n} \int_{B_\rho(0)} |Du|^2 \leq \beta$, and suppose $\rho^{-n} \int_{B_\rho(0) \setminus \{(x,y) : |x| < \rho/4\}} |u - \varphi|^2 < \eta_0$ and $\rho^{2-n} \int_{B_\rho(0)} |Du|^2 - \Theta_u(0) < \eta_0$. Then $\rho^{-n} \int_{B_{\rho/2}(0)} |u - \varphi|^2 < \zeta$ and*

$$\operatorname{sing} u \cap B_{\rho/2}(0) \subset \textit{ the } (\zeta\rho)\textit{-neighbourhood of } \{0\} \times \mathbb{R}^{n-3}.$$

Proof: By the argument in the first part of the proof of Lemma 3, for any given $\zeta_0 > 0$ we have that

$$\sum_{j=0}^{3} \rho^j |D^j u - D^j \varphi| \leq \zeta_0 \text{ on } B_{5\rho/8}(0) \setminus \{(x,y) : |x| \leq \tfrac{\rho}{2}\},$$

provided that $\eta_0 = \eta_0(\beta, N, n, \zeta_0)$ is small enough. But then since $\varphi_r = \varphi_{y^j} \equiv 0$ we have in particular that

$$r(|u_r| + |u_y|) \leq \zeta_0 \text{ on } B_{5\rho/8}(0) \setminus \{(x,y) : |x| \leq \tfrac{\rho}{2}\},$$

so (by choosing ζ_0 small enough) we can verify the hypotheses of Lemma 3 for any given $\zeta > 0$, provided $\eta_0 = \eta_0(\beta, \zeta, n, N)$ is sufficiently small. Then by Lemma 3 we have $\rho^{-n} \int_{B_{3\rho/4}(0)} |u - \widetilde\varphi|^2 < \zeta$ for suitable HCM $\widetilde\varphi(x,y) \equiv \widetilde\varphi(x,0)$ and hence $\rho^{-n} \int_{B_{3\rho/4}(0) \setminus \{(x,y) : |x| < \rho/4\}} |\varphi - \widetilde\varphi|^2 \leq C\zeta$ by the triangle inequality. Then, again using the triangle inequality, and also the fact that both φ, $\widetilde\varphi$ are homogeneous of degree zero, we conclude that $\rho^{-n} \int_{B_{3\rho/4}(0)} |u - \varphi|^2 < C\zeta$. By (iii) of Section 3.7 we now have the remaining conclusion with $C\zeta^{1/n}$ in place of ζ. Hence after a change of notation we have the required conclusions as stated. □

Proof of Theorem 1: Let $\zeta \in (0, \zeta_0]$ be given, where $\zeta_0 = \zeta_0(n, N, \beta)$ is as in Lemma 3, and let $\eta = \eta(N, n, \zeta, \beta) \in (0, 10^{-2n}]$ be as in Lemma 3. Then Lemma 3 implies that there is an HCM φ with

(1) $$\sup_{B_{7\rho/8}(0) \setminus \{(x,y) : |x| \leq \rho/16\}} \sum_{j=0}^{3} \rho^j |D^j u(r,y) - D^j \varphi|_{C^3} \leq \zeta,$$

4.4. Energy Estimates

and

(2) $$\sup_{B^+_{3\rho/4}\setminus\{(r,y):r<\rho/4\}} \left| \int_{S^2} (|\nabla^{S^2} u(r,y)|^2 - |\nabla^{S^2}\varphi|^2)\, d\omega \right| \le$$
$$\le C \left(\rho^{-n} \int_{B_\rho(0)\setminus\{(x,y):|x|<\rho/8\}} r^2(u_r^2 + u_y^2) \right)^{1/(2-\alpha)}.$$

Notice that by (1) and the estimates of Section 3.6 we then have

(3) $$\sup_{B_{3\rho/4}(0)\setminus\{(x,y):|x|<\rho/16\}} \sum_{j=0}^{3} \rho^j |D^j u| \le C\beta.$$

For each $y \in \mathbb{R}^{n-3}$ with $|y| < \frac{\rho}{2}$ we let

(4) $$\sigma_y = \sup(\{0\} \cup \{\sigma \in (0, \tfrac{\rho}{2}] : \sigma^{-n} \int_{B_\sigma(0,y)\setminus\{(x,y):|x|<\sigma/8\}} r^2(u_r^2 + u_y^2) \ge \eta\}).$$

Notice that, since $\rho^{-n} \int_{B_\rho(0)} r^2(u_r^2 + u_y^2) < \eta^2 \le 10^{-2n}\eta$, we have automatically

(5) $$\sup \sigma_y \le 10^{-2}\rho.$$

By the "five times" covering lemma (see e.g. [FH69] or [Si83a]) we can find a countable pairwise-disjoint collection $\{B_{4\sigma_{y_j}}(0, y_j)\}$ such that

(6) $$\bigcup_{|y|\le \rho/2,\, \sigma_y > 0} \overline{B}_{4\sigma_y}(0,y) \subset \bigcup_j \overline{B}_{20\sigma_{y_j}}(0, y_j).$$

Notice in particular that (by definition of σ_y) we have

(7) $$\sigma^{-n} \int_{B_\sigma(0,y)\setminus\{(x,y):|x|<\sigma/8\}} r^2(u_r^2 + u_y^2) < \eta$$

for each $\sigma \in (\sigma_y, \tfrac{\rho}{2}]$, and so by exactly the same reasoning (involving the first part of Lemma 3 and Section 3.6) that we used to conclude (1), (3) above, and keeping in mind that $\sigma^{2-n}\int_{B_\sigma(0,y)}|Du|^2 \le 2^n\beta$ by the monotonicity of Section 2.4, we deduce (taking a smaller $\eta = \eta(n, \beta, N, \zeta)$ if necessary) that $\sup_{B_{7\sigma/8}\setminus\{(x,y):|x|\le\sigma/16\}} \sum_{j=0}^{3} \sigma^j |D^j u| \le C\beta$. Hence by Lemma 2 we have, for each $y_0 \in \mathbb{R}^{n-3}$ with $|y_0| \le \tfrac{\rho}{2}$, and for all $\sigma \in (\sigma_{y_0}, \tfrac{\rho}{2}]$,

(8) $$\sup_{B^+_{3\sigma/4}(y_0)\setminus\{(r,y):r<\sigma/8\}} \left| \nabla_{r,y} \int_{S^2} |\nabla^{S^2} u(r,y)|^2 \right| \le$$
$$\le C\sigma^{-n} \int_{B_\sigma(0,y_0)\setminus\{(x,y):|x|<\sigma/16\}} r^2(u_r^2 + u_y^2).$$

We now want to define a Whitney-type cover for $B^+_{\rho/2}(0)$, as follows. For $j \ge 2$ let $B_{\rho/2^{j+2}}(0, z_{j,k})$, $k = 1, \ldots, Q_j$, be a maximal pairwise disjoint collection of balls of radius $\rho/2^{j+2}$ and centers $(0, z_{j,k}) \in B_{\rho/2}(0) \cap (\{0\} \times \mathbb{R}^{n-3})$. Then for $j \ge 2$

(9) $$\bigcup_{k=1}^{Q_j} B_{\rho/2^j}(0, z_{j,k}) \supset B_{\rho/2}(0) \cap \{(x,y) : |x| < \rho/2^{j+1}\},$$

and, for any point $(x,y) \in B_{\rho/2}(0)$,

(10) $\qquad \text{card}\{k : (x,y) \in B_{\rho/2^{j-5}}(0, z_{j,k})\} \leq C, \quad C = C(n).$

Next let $\Omega_{1,1} = B_{\rho/2}^+(0) \setminus \{(r,y) : r < \frac{\rho}{8}\}$, $\widehat{\Omega}_{1,1} = B_\rho^+(0) \setminus \{(r,y) : r < \frac{\rho}{16}\}$, and $Q_1 = 1$, and define for $j \geq 2$ and $k = 1, \ldots, Q_j$

(11) $\qquad \Omega_{j,k} = B_{\rho/2^j}^+(z_{j,k}) \setminus \{(r,y) : r < \rho/2^{j+2}\},$

and

(12) $\qquad \widehat{\Omega}_{j,k} = B_{\rho/2^{j-1}}^+(z_{j,k}) \setminus \{(r,y) : r < \rho/2^{j+3}\}.$

Notice that all points $(r,y) \in \widehat{\Omega}_{j,k}$ satisfy $\rho/2^{j+3} \leq r < \rho/2^{j-1}$, and in particular

$$\widehat{\Omega}_{j,k} \cap \widehat{\Omega}_{i,\ell} = \varnothing, \quad |i - j| \geq 4,$$

so it follows from (10) that

(13) $\qquad \forall (r,y) \in B_\rho^+(0), \text{ card}\{(j,k) : (r,y) \in \widehat{\Omega}_{j,k}\} \leq C, \quad C = C(n).$

Also, by (9)

(14) $$B_{\rho/2}^+ \subset B_{\rho/2}^+ \cap \left(\bigcup_{j=2}^{\infty} \{(r,y) : \rho/2^{j+2} \leq r < \rho/2^{j+1}\} \cup \{(r,y) : r \geq \tfrac{1}{8}\} \right)$$
$$\subset \bigcup_{j=1}^{\infty} \bigcup_{k=1}^{Q_j} \Omega_{j,k}.$$

Now, by (5), $\Omega_{1,1}$ intersects no $B_{\sigma_y}(y)$, while for each (j,k) such that $\Omega_{j,k}$ does not intersect $B_{\sigma_{z_{j,k}}}(z_{j,k})$ we must have $\rho/2^j > \sigma_{z_{j,k}}$. Thus, in any case, if $\Omega_{j,k}$ does not intersect $B_{\sigma_{z_{j,k}}}(z_{j,k})$ we can apply (8) with $\sigma = \rho/2^{j-1}$, $y_0 = z_{j,k}$ ($y_0 = 0$ in case $j = k = 1$), to deduce

(15) $$\int_{\Omega_{j,k}} r \left| \nabla_{r,y} \int_{S^2} |\nabla^{S^2} u(r,y)|^2 \, d\omega \right| r^2 dr dy \leq C \int_{\widehat{\Omega}_{j,k}} \int_{S^2} r^2 (u_r^2 + u_y^2) \, d\omega \, r^2 dr dy.$$

On the other hand if $\Omega_{j,k}$ does intersect $B_{\sigma_{z_{j,k}}}^+(z_{j,k})$ then $j \geq 2$ (by (5)) and $\sigma_{z_{j,k}} \geq 2^{-j-2}\rho$, hence $\Omega_{j,k} \subset B_{4\sigma_{j,k}}^+(z_{j,k}) \subset \cup_i \overline{B}_{20\sigma_i}^+(y_i)$. Hence by summing in (15) and using (13), (14), we conclude that

(16) $$\int_{B_{\rho/2}^+(0) \setminus (\cup_j B_{20\sigma_j}^+(y_j))} \left| r \nabla_{r,y} \int_{S^2} |\nabla^{S^2} u(r,y)|^2 \, d\omega \right| r^2 dr dy \leq C \int_{B_\rho(0)} r^2 (u_r^2 + u_y^2) \, dx dy.$$

4.4. Energy Estimates

Notice also that using the monotonicity of energy in Section 2.4 and the definition (4) of σ_y, we have that for each j

$$\sigma_{y_j}^{-n} \int_{B_{40\sigma_{y_j}}(0,y_j)} r^2 |Du|^2 \leq C \leq C\eta^{-1} \sigma_{y_j}^{-n} \int_{B_{\sigma_{y_j}}(0,y_j) \setminus \{(x,y) : |x| < \sigma_j/8\}} r^2 (u_r^2 + u_y^2).$$

Hence by summing on j, and using the disjointness of the $B_{\sigma_{y_j}}(0, y_j)$, we deduce that

(17) $$\sum_j (\sigma_{y_j}^n + \int_{B_{40\sigma_{y_j}}(0,y_j)} r^2 |Du|^2) \leq C \int_{B_\rho(0)} r^2 (u_r^2 + u_y^2).$$

Now we want to use the collection $\{B_{40\sigma_{y_j}}(0, y_j)\}$ to construct a cut-off function. For each j, let $\zeta_j : (0, \infty) \times \mathbb{R}^{n-3} \to [0, 1]$ be a C^∞ function with $\zeta_j(r, y) \equiv 1$ outside $B_{40\sigma_{y_j}}^+(y_j)$, $\zeta_j(r, y) \equiv 0$ in $B_{20\sigma_{y_j}}^+(y_j)$ and with

(18) $$\sup_{B_\rho^+} |\nabla \zeta_j| \leq \frac{C}{\sigma_{y_j}}.$$

Now evidently, since the $\{B_{4\sigma_j}(y_j)\}$ are pairwise disjoint, at most a finite subcollection of the $B_{40\sigma_{y_j}}(0, y_j)$ can intersect a given compact subset of $\mathbb{R}^n \setminus (\{0\} \times \mathbb{R}^{n-3})$, so we can define a smooth function $\zeta : (0, \infty) \times \mathbb{R}^{n-3} \to [0, 1]$ by

$$\zeta = \prod_j \zeta_j.$$

By construction $\zeta \equiv 0$ on $\cup_j \overline{B}_{20\sigma_{y_j}}^+(y_j) \supset \cup_{|y| \leq \frac{\rho}{2}, \sigma_y > 0} B_{4\sigma_y}^+(y)$. In particular if $0 < r_0 < \frac{\rho}{2}$, $|y| < \frac{\rho}{2}$, then $\zeta(r_0, y_0) > 0 \Rightarrow r_0 > \sigma_{y_0}$ and hence

(19) $$r_0^{-n} \int_{B_{r_0}(0,y_0) \setminus \{(x,y) : |x| < \sigma_0/8\}} r^2 (u_r^2 + u_y^2) \leq \eta,$$

which (since $\frac{4}{3} \cdot \frac{7}{8} = \frac{7}{6}$) guarantees by Lemma 3 and the estimates of Section 3.6 that u is smooth on each of the subsets $B_{7r_0/6}(0, y_0) \setminus \{(x,y) : |x| < \frac{r_0}{2}\}$, and hence in particular the function $(\int_{S^2} |\nabla^{S^2} u(r, y)|^2 \, d\omega - \int_{S^2} |\nabla^{S^2} \varphi|^2 \, d\omega)$ is smooth in a neighbourhood of (r_0, y_0). Thus

$$f(r, y) \equiv \zeta(r, y) \int_{S^2} (|\nabla^{S^2} u(r, y)|^2 - |\nabla^{S^2} \varphi|^2) \, d\omega$$

is a smooth function of $(r, y) \in B_\rho^+(0)$.

Next we note that since f is smooth on $\{(r, y) : r \in (0, \frac{\rho}{2}], |y| < \frac{\rho}{2}\}$, we can integrate by parts with respect to the r-variable giving

(20)
$$\int_{B_{\rho/2}^+(0)} |f| r^2 \, drdy \leq$$
$$\leq \int_{|y| < \rho/2, \, r < \rho/2} |f| r^2 \, drdy \leq$$
$$\leq C \int_{|y| < \rho/2, \, \rho/4 < r < \rho/2} |f| r^2 \, drdy + \tfrac{1}{3} \int_{|y| < \rho/2, \, r < \rho/2} \left| \frac{\partial f}{\partial r} \right| r^3 \, drdy.$$

We emphasize that this is valid even if f is not bounded near $r = 0$, because we can first prove (by integration by parts) an inequality as in (20) with $r_\varepsilon = \max\{r - \varepsilon, 0\}$ in place of r, and then let $\varepsilon \downarrow 0$. Since $\zeta \equiv 1$ on $B^+_{\rho/2} \setminus \cup_{j=1}^Q B^+_{40\sigma y_j}(y_j)$ and since $D_k \zeta = \sum_i (D_k \zeta_i) \prod_{j \neq i} \zeta_j$, we then obtain, in view of (16), (18), (5) and since $\sqrt{r^2 + |y|^2} \leq \frac{3\rho}{4}$ for $r, |y| < \frac{\rho}{2}$,

$$\int_{B^+_{\rho/2}(0)} \left| \int_{S^2} (|\nabla^{S^2} u(r,y)|^2 - |\nabla^{S^2} \varphi|^2 \, d\omega) \right| r^2 \, drdy \leq$$

$$\leq C \int_{B^+_{3\rho/4}(0) \setminus \{(r,y) : r < \rho/4\}} \left| \int_{S^2} (|\nabla^{S^2} u(r,y)|^2 - |\nabla^{S^2} \varphi|^2) \, d\omega \right| r^2 \, drdy +$$

$$+ C \int_{B^+_\rho(0)} r^2 (u_r^2 + u_y^2) + C \sum_j \left(\sigma_{y_j}^n + \int_{B_{40\sigma y_j}(0, y_j)} r^2 |Du|^2 \right).$$

In view of (2) and (17) this proves the theorem. □

4.5 L^2 estimates

Here we are going to use the energy estimates of the previous section together with the monotonicity identity of Section 2.4 (and some variants of it) to obtain L^2 estimates for u. These will be needed in the next section in proving decay properties of the deviation function introduced there.

u continues to denote a $W^{1,2}(\Omega; N)$ energy minimizer, and we assume $\overline{B_2(0)} \subset \Omega$ and that

(i) $\quad 0 \in \operatorname{sing} u, \quad \Theta_u(0) \geq \theta_0, \quad \int_{B_2(0)} |Du|^2 \leq \beta,$

where β is a given constant and $\theta_0 \in \{\Theta_\varphi(0) : \varphi \text{ is an HCM with } 0 < \Theta_\varphi(0) \leq \beta\}$. Notice that by monotonicity (see Section 2.4) this implies

(ii) $\quad \rho^{2-n} \int_{B_\rho(z)} |Du|^2 \leq C\beta, \quad \forall \rho \in (0, 1], \ |z| \leq 1.$

With θ_0 as in (i)

(iii) $\quad S_+ = \{z \in \overline{B_1(0)} : \Theta_u(z) \geq \theta_0\}.$

Let $\varepsilon \in (0, \frac{\delta}{2}]$ be for the moment arbitrary (we eventually choose $\delta \leq \delta_0(n, N, \beta)$ and $\varepsilon \leq \varepsilon_0(n, N, \beta, \delta)$). S_+ will be assumed to satisfy a weak ε-approximation property like that in Lemma 2 of Section 3.4 with $m = n - 3$; thus for each $\rho \in (0, 1]$ and each $z \in S_+$ we assume that

(iv) $\quad S_+ \cap B_\rho(z) \subset \text{ the } (\varepsilon \rho)\text{-neighbourhood of } L_{z,\rho},$

4.5. L^2 estimates

where $L_{z,\rho}$ is an $(n-3)$-dimensional affine space containing z. We henceforth fix these affine spaces $L_{z,\rho}$. We also here assume that

(v) $\quad \sup_{z \in S_+} \int_{B_2(0)} \frac{|R_z u_{R_z}|^2}{R_z^n} \leq \varepsilon, \quad \sup_{z \in S_+, \rho \leq 1} \rho^{2-n} \int_{B_\rho(z)} |Du|^2 \leq \theta_0 + \varepsilon, \quad \sup_{z \in S_+} \Theta_u(z) \leq \theta_0 + \varepsilon,$

where θ_0 is as in (i).

Remark: We show in Section 4.7 below that for every given $\varepsilon > 0$ and $w_0 \in \text{sing } u$ with $\Theta_u(w_0) = \theta_0 \in \{\Theta_\varphi(0) \in (0, \beta] : \varphi \text{ is an HCM}\}$ there is $\sigma > 0$ (depending on u, w_0, ε) such that all of the above conditions are satisfied, by Lemma 2 of Section 3.4 and the monotonicity of Section 2.4, with the rescaled function $u_{w_1,\sigma_1} \equiv u \circ \eta_{w_1,\sigma_1}$ in place of u for any $w_1 \in B_\sigma(w_0) \cap \{z : \Theta_u(z) \geq \Theta_u(w_0)\}$ and $\sigma_1 \leq \sigma$. These facts are of crucial importance in the eventual applicability of the results of the present section.

We also here suppose that $z_0 \in S_+$, $\rho \in (0, \frac{1}{4}]$, $\gamma \in (0, \frac{1}{2}]$, and that there exist points $z_1, \ldots, z_{n-3}$ in $S_+ \cap B_\rho(z_0)$ such that

(vi) $\quad \{z_j - z_0\}_{j=1,\ldots,n-3}$ are linearly independent

$$\text{and } \sum_{j=1}^{n-3} ((z_j - z_0) \cdot a)^2 \geq \gamma \rho^2 |a|^2 \ \forall a \in L,$$

where L is the $(n-3)$-dimensional linear space spanned by $z_1 - z_0, \ldots, z_{n-3} - z_0$. Notice that this says that the $z_j - z_0$ are in "uniformly general position", up to the factor γ, in $B_\rho(z_0) \cap L$.

The main result of this section is the following:

Theorem 1 *Suppose N is real-analytic. There is $\varepsilon_0 = \varepsilon_0(n, N, \beta) > 0$ such that if (i), (iv), (v), (vi) hold with $\varepsilon = \varepsilon_0$, then for all $z \in S_+$*

$$\int_{B_{\rho/8}(z_0)} \frac{|R_z u_{R_z}|^2}{R_z^n} \leq C\rho^{-2} s(z)^{2-n} \int_{B_\rho(z_0)} \sum_{j=0}^{n-3} |R_{z_j} u_{R_{z_j}}|^2 + C \left(\frac{\rho^{n-2}}{s(z)^{n-2}} \right)^{1-1/(2-\alpha)} \times$$

$$\times \left(\int_{B_\rho(z_0) \setminus \{(x,y) : \text{dist}((x,y), z_0 + L) < \rho/8\}} \left(\frac{1}{\rho^2 s(z)^{n-2}} \sum_{j=0}^{n-3} |R_{z_j} u_{R_{z_j}}|^2 + \frac{|R_z u_{R_z}|^2}{R_z^n} \right) \right)^{1/(2-\alpha)},$$

where $\alpha = \alpha(n, N, \beta) \in (0, 1)$, $C = C(n, N, \beta, \gamma)$, and $s(z) = \rho + |z - z_0|$.

We shall need the following three lemmas in the proof:

Lemma 1 *Suppose $L, z_0, \ldots, z_{n-3}$ are as in (v) (although here we do not need to assume that $z_j \in S_+$). Let U be open, $U \cap B_{2\rho}(z_0) \neq \emptyset$. Then for any $v \in W^{1,2}(U)$ we have*

$$C^{-1}(r_L^2 v_{r_L}^2 + \rho^2 |D_L v|^2) \leq \sum_{j=0}^{n-3} |R_{z_j} v_{R_{z_j}}|^2 \leq C(r_L^2 v_{r_L}^2 + \rho^2 |D_L v|^2) \text{ in } B_{2\rho}(z_0) \cap U,$$

where $C = C(n,\gamma)$, $r_L(w) \equiv \text{dist}(w, z_0 + L)$, $r_L v_{r_L}|_w \equiv (w - w') \cdot Dv$, with w' the nearest point projection of w onto $z_0 + L$, and where D_L means gradient parallel to L.

If $\sigma \in (0, \rho]$ and $\zeta_0, \ldots, \zeta_{n-3}$ are any other points in $B_\rho(z_0)$ with $\zeta_1, \ldots, \zeta_{n-3} \in B_\sigma(\zeta_0)$ and with

$$\sum_{j=1}^{n-3} ((\zeta_j - \zeta_0) \cdot a)^2 \geq \gamma \sigma^2 |a|^2, \quad a \in L,$$

then

$$C^{-1}(r_L^2 v_{r_L}^2 + \sigma^2 |D_L v|^2) - C \sum_{j=0}^{n-3} \text{dist}^2(\zeta_j, z_0 + L)|Dv|^2 \leq \sum_{j=0}^{n-3} |R_{\zeta_j} v_{R_{\zeta_j}}|^2 \leq$$

$$\leq C(r_L^2 v_{r_L}^2 + \sigma^2 |D_L v|^2) + C \sum_{j=0}^{n-3} \text{dist}^2(\zeta_j, z_0 + L)|Dv|^2 \text{ on } B_\sigma(\zeta_0) \cap U$$

for suitable $C = C(n, \gamma)$. In particular

$$\sum_{j=0}^{n-3} |R_{z_j} v_{R_{z_j}}|^2 \leq C \left(\frac{\rho}{\sigma}\right)^2 \sum_{j=0}^{n-3} \left(|R_{\zeta_j} v_{R_{\zeta_j}}|^2 + \text{dist}^2(\zeta_j, z_0 + L)|Dv|^2\right) \text{ on } B_\sigma(\zeta_0) \cap U.$$

Remark: Notice that the inequality $\sum_{j=1}^{n-3}((z_j - z_0) \cdot a)^2 \geq \gamma \rho^2 |a|^2, a \in L$, means that $z_0, \ldots, z_{n-3}$ must be in "uniformly general position" in $z_0 + L$ up to the factor γ; likewise the condition $\sum_{j=1}^{n-3}((\zeta_j - \zeta_0) \cdot a)^2 \geq \gamma \sigma^2 |a|^2, a \in L$, requires that the nearest point projections $\zeta_0', \ldots, \zeta_{n-3}'$ should be in such uniformly general position in $B_\sigma(\zeta_0')$.

Proof of Lemma 1: By definition

(1) $$R_{z_j} v_{R_{z_j}}|_w = (w - z_j) \cdot Dv$$
$$= (w - w') \cdot Dv + (w' - z_j) \cdot Dv$$
$$= r_L v_{r_L} + (w' - z_j) \cdot D_L v$$

so in particular

$$R_{z_j} v_{R_{z_j}} - R_{z_0} v_{R_{z_0}} = (z_0 - z_j) \cdot D_L v,$$

and by the hypothesis we then have that on U

(2) $$\gamma \rho^2 |D_L v|^2 \leq \sum_{j=0}^{n-3} (R_{z_j} v_{R_{z_j}} - R_{z_0} v_{R_{z_0}})^2.$$

On the other hand using (1) with $j = 0$ we also have on $U \cap B_{2\rho}(z_0)$ that

(3) $$r_L^2 v_{r_L}^2 \leq 2(R_{z_0} v_{R_{z_0}})^2 + C\rho^2 |D_L v|^2.$$

4.5. L^2 estimates

Combining (2) and (3) we then have, with $C = C(\gamma, n)$,

$$r_L^2 v_{r_L}^2 + \rho^2 |D_L v|^2 \leq C \sum_{j=0}^{n-3} |R_{z_j} v_{R_{z_j}}|^2$$

as claimed. Notice that the reverse inequality $C^{-1} \sum_{j=0}^{n-3} |R_{z_j} v_{R_{z_j}}|^2 \leq (r_L v_{r_L})^2 + \rho^2 |D_L v|^2$ follows directly from (1) on $B_{2\rho}(z_0) \cap U$.

Next notice (Cf. (1) above) that at any point $w \in B_\sigma(\zeta_0)$

(4) $\quad R_{\zeta_j} v_{R_{\zeta_j}} = (w - \zeta_j) \cdot Dv$

$\qquad = (w - w') \cdot Dv + (w' - \zeta_j') \cdot Dv + (\zeta_j' - \zeta_j) \cdot Dv$

$\qquad = r_L v_{r_L} + (w - \zeta_j) \cdot D_L v + (\zeta_j' - \zeta_j) \cdot Dv.$

Taking differences in (4) we see that

$$(\zeta_j - \zeta_0) \cdot D_L v = -R_{\zeta_j} v_{R_{\zeta_j}} + R_{\zeta_0} v_{R_{\zeta_0}} + (\zeta_j' - \zeta_j) \cdot Dv - (\zeta_0' - \zeta_0) \cdot Dv.$$

Since $|\zeta_j' - \zeta_j| = \mathrm{dist}(\zeta_j, z_0 + L)$, by using the given hypothesis on the ζ_j we then see that on U

(5) $\quad \sigma^2 |D_L v|^2 \leq C \sum_{j=0}^{n-3} (R_{\zeta_j} v_{R_{\zeta_j}})^2 + C \sum_{j=0}^{n-3} \mathrm{dist}^2(\zeta_j, z_0 + L) |Dv|^2.$

Going back to (4) again we then also conclude that on $U \cap B_\sigma(\zeta_0)$

$$r_L^2 v_{r_L}^2 \leq C \sum_{j=0}^{n-3} (R_{\zeta_j} v_{R_{\zeta_j}})^2 + C \sum_{j=0}^{n-3} \mathrm{dist}^2(\zeta_j, z_0 + L) |Dv|^2,$$

which proves the required upper inequality for $r_L^2 v_{r_L}^2 + \sigma^2 |D_L v|^2$. The reverse inequality follows directly from (4) and the triangle inequality.

The final inequality of the lemma is simply a matter of combining two of the previous inequalities, so this completes the proof of the lemma. $\square$

In the proof of Theorem 1 we shall want to apply the main energy estimate established in Theorem 1 of Section 4.4, and this requires that we check the hypothesis that u is L^2-sufficiently close to some HCM φ with $S(\varphi) = \{0\} \times \mathbb{R}^{n-3}$ in the appropriate ball. For this we need the following lemma.

Lemma 2 *For any given $\zeta > 0$ there is $\varepsilon_0 = \varepsilon_0(N, n, \beta, \zeta) > 0$ such that if (i), (iv), (v), (vi) hold with $\varepsilon \leq \varepsilon_0$, then*

$$\rho^{-n} \int_{B_\rho(z_0)} (r_L^2 u_{r_L}^2 + \rho^2 |D_L u|^2) \leq C\varepsilon,$$

where the notation is as in Lemma 1, and

$$\rho^{-n}\int_{B_{\rho/2}(z_0)}|u-\varphi_{z_0}|^2\leq\zeta,\qquad \sup_{B_{2\rho/3}(z_0)\setminus\{(x,y):r_L\leq\rho/16\}}\sum_{j=0}^{3}\rho^j|D^ju-D^j\varphi_{z_0}|_{C^3}\leq\zeta$$

for some HCM φ with $S(\varphi)=L$, $\theta_0-\zeta\leq\Theta_\varphi(0)\leq\theta_0+\zeta$ ($\Theta_\varphi(0)=\theta_0$ if N is real analytic). Here $\varphi_{z_0}(x,y)\equiv\varphi((x,y)-z_0)$. Furthermore there is $\varepsilon_0=\varepsilon_0(N,n,\beta)>0$ such that if (i), (iv), (v), (vi) hold with $\varepsilon\leq\varepsilon_0$ then for all $z\in S_+$

$$\mathrm{dist}^2(z,z_0+L)\leq$$
$$\leq C\rho^{2-n}\int_{\{(x,y)\in B_{3\rho/4}(z_0):r_L\geq\rho/4\}}(r_L^2 u_{r_L}^2+(\rho+|z-z_0|)^2|D_L u|^2+|R_z u_{R_z}|^2)\leq$$
$$\leq C\varepsilon\rho^{2-n}(\rho+|z-z_0|)^n.$$

Remark: It is not assumed that $|z-z_0|$ is small here; z_0, z are unrelated points in S_+.

Proof of Lemma 2: Evidently we can assume without loss of generality that L in (vi) is $\{0\}\times\mathbb{R}^{n-3}$. To prove the first inequality, notice that by Lemma 1 above we have

$$r_0^2 u_{r_0}^2+\rho^2 u_y^2\leq C\sum_{j=0}^{n-3}|R_{z_j}u_{R_{z_j}}|^2,$$

where $r_0=|x-\xi_{z_0}|$, $r_0 u_{r_0}=(x-\xi_{z_0})\cdot u_x$, $z_0=(\xi_{z_0},\eta_{z_0})$. Integrating this inequality over the ball $B_\rho(z_0)$ and noting that (v) implies

(1) $$\rho^{-n}\int_{B_\rho(z_0)}\sum_{j=0}^{n-3}|R_{z_j}u_{R_{z_j}}|^2\leq C\varepsilon,$$

we then have the first inequality as claimed.

In view of the first inequality, the first part of Lemma 3 of Section 4.4 guarantees that the second and third inequalities of the lemma hold for some HCM φ with $\varphi(x,y)\equiv\varphi(x,0)$ and

(2) $$\theta_0-\zeta\leq\Theta_\varphi(0)\leq\theta_0+\zeta,$$

and

(3) $$\sup_{B_{2\rho/3}(z_0)\setminus\{(x,y):|x-\xi_{z_0}|\leq\rho/16\}}\sum_{j=0}^{3}\rho^j|D^ju-D^j\varphi_{z_0}|_{C^3}\leq\zeta.$$

In case N is real-analytic, we agree that ζ is chosen smaller than the minimum distance between distinct elements of $\{\Theta_\varphi(0):\varphi$ is an HCM with $\Theta_\varphi(0)\leq\beta\}$. Then (2) gives $\Theta_\varphi(0)=\theta_0$ in case N is real-analytic. We next claim that (for ζ small enough in (3))

(4) $$|\xi|^2\leq C\rho^{2-n}\int_{|y-\eta_{z_0}|<\rho/2,\,\rho/4<r_0<\rho/2}(\xi\cdot u_x)^2,$$

4.5. L^2 estimates

where $C = C(n, N, \beta)$ is fixed (independent of ξ, u), provided $\varepsilon_0 = \varepsilon_0(n, N, \beta) > 0$ is small enough. Indeed otherwise by (2) and (3), after rescaling and translating so that $\rho = 1$ and $z_0 = 0$, we would have a sequence $u_k \in W^{1,2}(B_1(0); N)$, with $0 \in \text{sing } u_k$, φ_k HCM's with $S(\varphi_k) = \{0\} \times \mathbb{R}^{n-3}$, and points $\xi_k \in S^2$ such that $\int_{B_1(0)} |Du_k|^2 \leq \beta$, $\int_{B_1(0)} |D\varphi_k|^2 \leq C\beta$, $\sup_{B_{2/3}(0) \setminus \{(x,y) : |x| \leq 1/16\}} \sum_{j=0}^{3} \rho^j |D^j u_k - D^j \varphi_k|_{C^3} \to 0$ as $k \to \infty$, and

(5) $\qquad \xi_k \to \xi \in S^2, \quad \int_{|y| < \frac{1}{2}, \frac{1}{4} < |x| < \frac{1}{2}} (\xi_k \cdot D_x u_k)^2 \to 0.$

Notice we also have
$$\liminf_{k \to \infty} \Theta_{\varphi_k}(0) > 0$$
by virtue of Corollary 2 of Section 2.10. Using the compactness theorem we can assume that $\varphi_k \to \varphi$ locally in $\mathbb{R}^n$, $u_k \to \varphi$ in $B_{2/3}(0) \setminus \{(x,y) : |x| \leq \frac{1}{16}\}$ and that $\xi \cdot \varphi_x \equiv 0$. But this implies that $\varphi((x,y) + \lambda(\xi, 0)) \equiv \varphi(x, y)$ for every $\lambda \in \mathbb{R}$, so $\text{sing } \varphi$ contains the ray in the direction of $(\xi, 0)$, contradicting the fact that $\text{sing } \varphi = \{0\} \times \mathbb{R}^{n-3}$. (Notice that φ is not constant because $\Theta_\varphi(0) > 0$ by upper-semicontinuity as in Section 2.11.) Thus (4) is established.

On the other hand we have, using the notation $z_0 = (\xi_{z_0}, \eta_{z_0})$, $z = (\xi_z, \eta_z)$,
$$(\xi_{z_0} - \xi_z) \cdot u_x = R_z u_{R_z} - r_0 u_{r_0} - (y - \eta_z) \cdot u_y,$$
and hence
$$|(\xi_{z_0} - \xi_z) \cdot u_x|^2 \leq 3(R_z u_{R_z})^2 + 3(r_0 u_{r_0})^2 + 3((y - \eta_z) \cdot u_y)^2.$$

By integrating this identity and using (4) with $\xi = \xi_{z_0} - \xi_z$ we have
$$|\xi_z - \xi_{z_0}|^2 \leq C\rho^{2-n} \int_{|y - \eta_{z_0}| < \rho/2, \, \rho/4 < r_0 < \rho/2} \left(r_0^2 u_{r_0}^2 + (\rho + |z - z_0|)^2 u_y^2 + |R_z u_{R_z}|^2 \right)$$
$$\leq C\varepsilon \rho^{2-n} (\rho + |z - z_0|)^n$$
by (v) and Lemma 1, as claimed. $\square$

The third lemma is as follows:

Lemma 3 *For any HCM φ with $S(\varphi) = \{0\} \times \mathbb{R}^{n-3}$ and any Lipschitz ψ on B_ρ^+ with $\psi(r, y) \equiv 0$ for $r^2 + |y|^2 = \rho^2$, we have the identity*
$$\int_{B_\rho} (|Du|^2 - |D\varphi|^2 + 2|u_y|^2)\psi = -\int_{B_\rho} r(|Du|^2 - |D\varphi|^2)\psi_r +$$
$$+ 2\int_{B_\rho} r|u_r|^2 \psi_r + 2\int_{B_\rho} r \sum_{j=1}^{n-3} u_r \cdot u_{y^j} \psi_{y^j}.$$

Proof: We begin by recalling the identity (v) of Section 2.2, which is valid for any Lipschitz $\zeta = (\zeta^1, \ldots, \zeta^n) : \overline{B}_\rho \to \mathbb{R}^n$ with $\zeta = 0$ on ∂B_ρ. Taking $\zeta = \psi(r,y)(x,0)$ (where $r = |x|$) in this identity, we thus obtain

$$\int_{B_\rho} \sum_{i=1}^{n}\sum_{j=1}^{3}(\delta_{ij}|Du|^2 - 2D_i u \cdot D_j u)\delta_{ij}\psi =$$

$$= -\int_{B_\rho} \sum_{i=1}^{n}\sum_{j=1}^{3}(\delta_{ij}|Du|^2 - 2D_i u \cdot D_j u)x^j D_i[\psi(r,y)].$$

Since $D_i[\psi(r,y)] = r^{-1}x^i\psi_r$ if $i \leq 3$ and $D_i[\psi(r,y)] = D_{y^{i-3}}\psi$ if $i = 4,\ldots,n$, we thus have

$$\int_{B_\rho}(|Du|^2 + 2|u_y|^2)\psi = -\int_{B_\rho} r|Du|^2\psi_r + 2\int_{B_\rho} r|u_r|^2\psi_r + 2\int_{B_\rho} r\sum_{j=1}^{n-3} u_r \cdot u_{y^j}\psi_{y^j}.$$

Now on the other hand by cylindrical homogeneity of φ (which guarantees that $|D\varphi|^2(x,y) = r^{-2}|D\varphi|^2(\omega,0)$) and the fact that the volume element of $\mathbb{R}^n$ is in the chosen coordinates $r^2\,d\omega dr dy$, we have by integration by parts in the r-variable that

$$-\int_{B_\rho}\psi|D\varphi|^2 = \int_{B_\rho} r\psi_r|D\varphi|^2,$$

so by adding this to the previous inequality we conclude the identity claimed in the statement of the lemma. $\square$

Proof of Theorem 1: By rotating if necessary, we may assume that the subspace L of (vi) is $\{0\} \times \mathbb{R}^{n-3}$. If $z_0 = (\xi_0, \eta_0)$, then we have $z_j = (\xi_0, \eta_j)$ for $j = 1,\ldots, n-3$. By the monotonicity identity (ii) of Section 2.4 we have, for any HCM φ such that $\Theta_\varphi(0) = \theta_0$ and $S(\varphi) = \{0\} \times \mathbb{R}^{n-3}$, and for any $z \in S_+ \cap B_\rho(z_0)$

$$2\int_{B_\rho(z)} \frac{|R_z u_{R_z}|^2}{R_z^n} \leq \frac{\rho^{3-n}}{n-2}\int_{\partial B_\rho(z)}(|Du|^2 - |D\varphi_z|^2),$$

where we used the fact that $\Theta_u(z) \geq \Theta_\varphi(0) = \theta_0$ and where $\varphi_z(x,y) \equiv \varphi((x,y)-z) \equiv \varphi_0(x - \xi_z)$. Let $\psi : \mathbb{R} \to [0,1]$ satisfy $\psi(t) \equiv 0$ for $t \geq \rho$, $\psi(t) \equiv 1$ for $t \leq \frac{(1+\theta)}{2}\rho$, $\psi' \leq 0$ everywhere, and $|\psi'(t)| \leq C(\theta)\rho^{-1}$. Multiplying each side of this inequality by $\psi(\rho)$ and integrating over $[\theta\rho, \rho]$ we get for any $\theta \in (0,1)$ that

(1) $$2\int_{B_{\theta\rho}(z)} \frac{|R_z u_{R_z}|^2}{R_z^n} \leq C\rho^{2-n}\int_{B_\rho(z)} \psi(R_z)(|Du|^2 - |D\varphi_z|^2).$$

On the other hand the identity of Lemma 3 above implies

$$\int_{B_\rho(z)} \psi(R_z)(|Du|^2 - |D\varphi_z|^2 + 2|u_y|^2) \leq \int_{B_\rho(z)} r_z^2 R_z^{-1}|\psi'(R_z)|(|Du|^2 - |D\varphi_z|^2) +$$

$$+ 2\int_{B_\rho(z)} r_z^2 R_z^{-1}|\psi'(R_z)|u_{r_z}^2 + 2\int_{B_\rho(z)} \sum_{j=1}^{n-3}\left(\frac{y^j}{R_z}\right) r_z u_{r_z} \cdot u_{y^j}\psi'(R_z)$$

4.5. L^2 estimates

where C depends on θ. Replacing ψ by ψ^2 and using the Cauchy-Schwarz inequality we have

(2)
$$\int_{B_\rho(z)} \psi^2(R_z)(|Du|^2 - |D\varphi_z|^2) \leq 4 \int_{B_\rho(z)} r_z^2 R_z^{-1} \psi(R_z)|\psi'(R_z)|(|Du|^2 - |D\varphi_z|^2) +$$
$$+ 4 \int_{B_\rho(z)} (R_z^{-1}\psi(R_z)|\psi'(R_z)| + (\psi'(R_z))^2) r_z^2 u_{r_z}^2,$$

and hence by combining (1) and (2) we have

(3)
$$\int_{B_{\theta\rho}(z)} \frac{|R_z u_{R_z}|^2}{R_z^n} \leq C\rho^{-n} \int_{B_\rho(z)} r_z^2 u_{r_z}^2 +$$
$$+ C\rho^{-n} \int_{B_\rho^+} \left| \int_{S^2} r^2(|Du^z|^2 - |D\varphi|^2) \, d\omega \right| r^2 \, dr \, dy,$$

where $u^z(x, y) \equiv u((x, y) + z)$. Now by (v), Lemma 1 and Lemma 2 (with 2ρ in place of ρ) we have

(4)
$$|\xi_z - \xi_{z_0}|^2 \leq C\rho^{2-n} \int_{B_{2\rho}(z_0) \setminus \{(x,y) : |x - \xi_{z_0}| \leq \rho/2\}} (r_0^2 u_{r_0}^2 + (\rho + |z - z_0|)^2 u_y^2 + |R_z u_{R_z}|^2)$$
$$\leq C\varepsilon \rho^{2-n}(\rho + |z - z_0|)^n.$$

(Notice that for the present we need this only for the case $z \in S_+ \cap B_\rho(z_0)$, but in fact Lemma 2 shows that it is valid for all $z \in S_+$.) Since $r_z^2 u_{r_z}^2 = ((x - \xi_z) \cdot u_x)^2 = ((x - \xi_{z_0}) \cdot u_x + ((\xi_z - \xi_{z_0}) \cdot u_x)^2 \leq 2r_0^2 u_{r_0}^2 + |\xi_z - \xi_{z_0}|^2 u_x^2$, we then have by (4), (v) and the first part of Lemma 2 that

(5)
$$\int_{B_{2\rho}(z)} (r_z^2 u_{r_z}^2 + \rho^2 u_y^2) \leq$$
$$\leq C \int_{B_{2\rho}(z)} (r_0^2 u_{r_0}^2 + \rho^2 u_y^2) + \int_{B_{2\rho}(z_0) \setminus \{(x,y) : |x - \xi_{z_0}| < \rho\}} (r_0^2 u_{r_0}^2 + \rho^2 u_y^2 + |R_z u_{R_z}|^2)$$
$$\leq C \int_{B_{5\rho/2}(z_0)} (r_0^2 u_{r_0}^2 + \rho^2 u_y^2) + \int_{B_{2\rho}(z_0) \setminus \{(x,y) : |x - \xi_{z_0}| < \rho/2\}} (r_0^2 u_{r_0}^2 + \rho^2 u_y^2 + |R_z u_{R_z}|^2)$$
$$\leq C\varepsilon \rho^n,$$

assuming $z \in S_+ \cap B_{\rho/2}(z_0)$. In particular with ε small enough we can apply the main energy estimate Theorem 1 of Section 4.4 with 2ρ in place of ρ on the right side of (3), thus obtaining (after selecting $\theta = \frac{5}{6}$)

(6)
$$\int_{B_{5\rho/6}(z)} \frac{|R_z u_{R_z}|^2}{R_z^n} \leq C\rho^{-n} \int_{B_{2\rho}(z)} r_z^2(u_{r_z}^2 + u_y^2) +$$
$$+ C \left(\rho^{-n} \int_{B_{2\rho}(z) \setminus \{(x,y) : |x - \xi_z| < \rho\}} r_z^2(u_{r_z}^2 + u_y^2) \right)^{1/(2-\alpha)}.$$

Using (5) again on the right of (6) and also replacing ρ by $\frac{3\rho}{4}$ we obtain for all $z \in B_{3\rho/8}(z_0)$

(7) $\displaystyle\int_{B_{\rho/4}(z_0)} \frac{|R_z u_{R_z}|^2}{R_z^n} \leq +C\rho^{-n} \int_{B_{2\rho}(z_0)} (r_0^2 u_{r_0}^2 + \rho^2 u_y^2) +$

$+ C\left(\rho^{-n} \displaystyle\int_{B_{2\rho}(z_0)\setminus\{(x,y):|x-\xi_{z_0}|<\rho/4\}} (r_0^2 u_{r_0}^2 + \rho^2 u_y^2) + |R_z u_{R_z}|^2\right)^{1/(2-\alpha)}.$

Notice that here we used the fact that $|\xi_z - \xi_{z_0}| \leq \frac{\rho}{4}$ for $z \in S_+ \cap B_\rho(z_0)$, by (4), and we also used the inclusions $B_{3\rho/2}(z) \subset B_{2\rho}(z_0)$, $B_{\rho/4}(z_0) \subset B_{5\rho/8}(z)$ for $z \in B_{3\rho/8}(z_0)$.

Now we want to consider $|z - z_0| \geq \frac{3\rho}{8}$. Notice that then, with $z = (\xi_z, \eta_z)$, we have

$|R_z u_{R_z}|^2 = ((x - \xi_z) \cdot u_x + (y - \eta_z) \cdot u_y)^2 \leq C(r_0^2 u_{r_0}^2 + |y - \eta_z|^2 u_y^2 + |\xi_z - \xi_{z_0}|^2 |u_x|^2).$

By integrating this inequality over the ball $B_{\rho/4}(z_0)$ (keeping in mind that we have the bound $\rho^{2-n} \int_{B_\rho(z_0)} u_x^2 \leq C\beta$ by (ii)), and using (4) (with $\frac{\rho}{2}$ in place of ρ), we obtain

$\displaystyle\int_{B_{\rho/4}(z_0)} |R_z u_{R_z}|^2 \leq C \int_{B_\rho(z_0)} (r_0^2 u_{r_0}^2 + s^2(z) u_y^2) + C \int_{B_\rho(z_0)\setminus\{(x,y):|x-\xi_{z_0}|\leq \rho/4\}} |R_z u_{R_z}|^2,$

where $s(z) = \rho + |z - z_0|$. Notice that since $|z - z_0| \geq \frac{3\rho}{8}$ this implies

$\displaystyle\int_{B_{\rho/4}(z_0)} \frac{|R_z u_{R_z}|^2}{R_z^n} \leq$

$\leq Cs^{-n}(z) \displaystyle\int_{B_\rho(z_0)} (r_0^2 u_{r_0}^2 + s^2(z) u_y^2) + Cs^{-n}(z) \int_{B_\rho(z_0)\setminus\{(x,y):|x-\xi_{z_0}|\leq \rho/4\}} |R_z u_{R_z}|^2$

$\leq Cs^{-n}(z) \displaystyle\int_{B_\rho(z_0)} (r_0^2 u_{r_0}^2 + s^2(z) u_y^2) +$

$+ C\left(\dfrac{\rho^{n-2}}{s(z)^{n-2}}\right)^{1-(1/(2-\alpha))} \left(s^{-n}(z) \displaystyle\int_{B_\rho(z_0)\setminus\{(x,y):|x-\xi_{z_0}|\leq \rho/4\}} |R_z u_{R_z}|^2\right)^{1/(2-\alpha)},$

where we used the fact that $\int_{B_\rho(z_0)} |R_z u_{R_z}|^2 \leq C\rho^{n-2} s^2(z)$ by (ii). Using this in case $|z - z_0| \geq \frac{3\rho}{8}$ and using (7) in case $|z - z_0| < \frac{3\rho}{8}$, we thus have

(8) $\displaystyle\int_{B_{\rho/4}(z_0)} \frac{|R_z u_{R_z}|^2}{R_z^n} \leq Cs^{-n}(z) \int_{B_{2\rho}(z_0)} (r_0^2 u_{r_0}^2 + s^2(z) u_y^2) +$

$+ C(\dfrac{\rho^{n-2}}{s(z)^{n-2}})^{1-\frac{1}{2-\alpha}} \left(\displaystyle\int_{B_{2\rho}(z_0)\setminus\{(x,y):|x-\xi_{z_0}|\leq \rho/4\}} (\dfrac{r_0^2 u_{r_0}^2 + s^2(z) u_y^2}{s^n(z)} + \dfrac{|R_z u_{R_z}|^2}{R_z^n})\right)^{\frac{1}{2-\alpha}}$

for every $z \in S_+$.

The proof is now completed by using the first conclusion of Lemma 1 (with $L = \{0\} \times \mathbb{R}^{n-3}$) in each of the integrals on the right side of this inequality and then replacing ρ by $\frac{\rho}{2}$. □

4.6 The deviation function ψ

Here we use the gap measures of Section 4.3 in order to construct a certain deviation function ψ, where $\psi(x,y)$ is the mean over $z \in S_+$ (S_+ as in Section 4.5) of the quantity $|(x,y) - z|^{-n}|((x,y)-z) \cdot Du(x,y)|^2$ (which appears on the left of the main inequality in Theorem 1 of Section 4.5) with respect to a gap measure constructed as in Section 4.3, with S_+ in place of S.

We continue to assume the hypotheses (i) (hence (ii)) and (iii), (iv), (v) of Section 4.5.

Let $\rho \in (0, \frac{1}{4}]$, $\delta \in (0, \frac{1}{16})$ (smaller than the $\delta_0(n)$ of Lemma 1 of Section 4.3), and let $S_\rho^+, T_\rho^+, \mu^+$ corresponding to S_ρ, T_ρ, μ of Section 4.3 with S_+ in place of S. Notice that by definition of T_ρ^+, we have $\mathrm{dist}(z, z_1 - L_{0,1}) \leq C\delta\rho$ for $z_1 \in T_\rho \cap S_+$ and $z \in S_+ \cap B_\rho(z_1)$. Henceforth we assume without loss of generality that $L_{0,1} = \{0\} \times \mathbb{R}^{n-3}$, as we did in the proof of Lemma 1 of Section 4.3, so this gives

(i) $\quad |\xi_z - \xi_{z_1}| \leq C\delta\rho, \quad z_1 = (\xi_{z_1}, \eta_{z_1}) \in T_\rho \cap S_+, \ z = (\xi_z, \eta_z) \in B_\rho(z_1) \cap S_+, \ \rho \in (0, \frac{1}{8}].$

Now define the deviation function ψ by

(ii) $$\psi(x,y) = \int_{S_+} \left. \frac{|R_z u_{R_z}|^2}{R_z^n} \right|_{(x,y)} d\mu^+(z).$$

Notice that for given $(x,y) \in \overline{B}_1(0) \setminus \mathrm{sing}\, u$, the integrand in (ii) is an analytic function of $z \in S_+$, so ψ is certainly well-defined on $\overline{B}_1(0) \setminus \mathrm{sing}\, u$.

Notice $\psi(x,y) \equiv 0$ if u itself is an HCM with $\Theta_u(0) = \theta_0$, and in general $\int_{B_1(0)} \psi$ measures the deviation (in an L^2-sense) of u away from such an HCM. Notice also that by (v) of Section 4.5 we have

$$\int_{B_1(0)} \psi(x,y)\, dxdy \leq C\varepsilon.$$

The main result concerning the deviation function is the following:

Theorem 1 *Suppose N is real-analytic and $\beta > 0$. There is $\delta_0 = \delta_0(n, N, \beta) > 0$ such that the following holds. If (i)–(v) of Section 4.5 hold, if $\delta \leq \delta_0$, and if $\varepsilon = \varepsilon(n, N, \beta, \delta) \leq \frac{\delta}{16}$ is small enough, then for any $\rho \in (0, \frac{1}{16}]$ we have the estimate*

$$\int_{T_{\theta\rho}^+} \psi \leq C \left(\int_{T_\rho^+ \setminus T_{\theta\rho}^+} \psi \right)^{1/(2-\alpha)},$$

where $\alpha = \alpha(n, N, \beta) \in (0,1)$ and $\theta = \theta(n, N, \beta) \in (0, \frac{1}{32}]$.

Proof: The proof is based on the L^2 estimates of the previous section. As mentioned above, we assume

(1) $$L_{0,1} = \{0\} \times \mathbb{R}^{n-3}.$$

Take $\rho \in (0, \frac{1}{8}]$. If $T_\rho^+ = \emptyset$ then we have nothing further to prove, so assume that $T_\rho^+ \neq \emptyset$, and take an arbitrary point $w_0 \in T_\rho^+ \cap S_+$. We have by definition of $T_\rho^+ (\subset T_{2\rho}^+)$ that there is a point $\widetilde{w}_0 \in B_\rho(w_0) \cap \widetilde{S}_+$ such that

(2) $$B_{2\rho}(\widetilde{w}_0) \cap S_+ \subset \{w : \operatorname{dist}(w, \widetilde{w}_0 + \{0\} \times \mathbb{R}^{n-3}) < 2\delta\rho\}$$

and

(3) $$B_{3\delta\rho}(w) \cap S_+ \neq \emptyset \quad \forall w \in (\widetilde{w}_0 + \{0\} \times \mathbb{R}^{n-3}) \cap B_{2\rho}(\widetilde{w}_0).$$

Thus it follows that

(4) $$B_\rho(w_0) \cap S_+ \subset \text{ the } (6\delta\rho)\text{-neighbourhood of } w_0 + \{0\} \times \mathbb{R}^{n-3},$$

and

(5) $$B_{6\delta\rho}(w) \cap S_+ \neq \emptyset \quad \forall w \in (w_0 + \{0\} \times \mathbb{R}^{n-3}) \cap B_\rho(w_0).$$

Also since any $w \in B_\rho(w_0) \cap S_+$ is in $T_{2\rho}^+ \cap S_+$, we know by Lemma 1 of Section 4.3 that

(6) $$C^{-1} \sigma^{n-3} \leq \mu^+(B_\sigma(w) \cap S_+) \leq C \sigma^{n-3}, \quad \forall \sigma \in [4\delta^{1/2}\rho, \tfrac{1}{16}]$$

and for any $w \in (w_0 + \{0\} \times \mathbb{R}^{n-3}) \cap B_\rho(w_0)$, where $C = C(n)$. Now let $w_1, \ldots, w_{n-3}$ be any points in $(w_0 + \{0\} \times \mathbb{R}^{n-3}) \cap B_\rho(w_0)$ such that

(7) $$\sum_{j=1}^{n-3} ((w_j - w_0) \cdot a)^2 \geq \frac{\rho^2}{2} |a|^2 \quad \forall a \in \{0\} \times \mathbb{R}^{n-3}.$$

Let $\theta \in [8\delta^{1/2}, \tfrac{1}{64}]$ be for the moment arbitrary. (We choose $\theta = \theta(n, N, \beta)$ below; notice that since we require $8\delta^{1/2} \leq \theta$, this also requires that δ be chosen small depending on n, N, β. In fact we are going to complete the proof with $\delta_0 = \theta^2(n, N, \beta)/64$.) In view of (5) and (6), for each $j \in \{0, \ldots, n-3\}$ we can select points $z_j \in B_{\rho/32}(w_j) \cap S_+$ such that

(8) $$\int_{B_\rho(w_0) \setminus \{(x,y) : |x - \xi_{w_0}| < \theta\rho/8\}} \frac{|R_{z_j} u_{R_{z_j}}|^2}{R_{z_j}^n} \leq$$

$$\leq C \mu^+(B_{\rho/8}(w_j))^{-1} \int_{B_{\rho/8}(w_j)} \int_{B_\rho(w_0) \setminus \{(x,y) : |x - \xi_{w_0}| < \theta\rho/8\}} \frac{|R_z u_{R_z}|^2}{R_z^n} \, dx dy \, d\mu^+(z)$$

$$\leq C \rho^{-(n-3)} \int_{B_\rho(w_0) \setminus \{(x,y) : |x - \xi_{w_0}| < \theta\rho/8\}} \psi(x, y) \, dx dy.$$

4.6. The deviation function ψ

Notice that here we use the general principle that for any Borel set $U \times V \subset B_\rho(w_0) \times S_+$ and any $\Gamma > 0$ we have

(9)
$$\int_U \frac{|R_\zeta u_{R_\zeta}|^2}{R_\zeta^n} \leq \Gamma \mu^+(V)^{-1} \int_V \int_U \frac{|R_z u_{R_z}|^2}{R_z^n}\, dxdy\, d\mu^+(z)$$

$$= \Gamma \mu^+(V)^{-1} \int_U \int_V \frac{|R_z u_{R_z}|^2}{R_z^n}\, d\mu^+(z)\, dxdy$$

$$\leq \Gamma \mu^+(V)^{-1} \int_U \psi(x,y)\, dxdy$$

for all $\zeta \in V$ except for a Borel set $E \subset V$ with $\mu^+(E) \leq \Gamma^{-1}\mu^+(V)$. (Notice that this implies that if U_1, U_2 are two subsets of $B_\rho(w_0)$ and if $\Gamma > 2$ then there exists at least one point $\zeta \in V$ such that we simultaneously have (9) with each of the choices $U = U_1$, $U = U_2$.)

Also, since $|z_j - w_j| \leq \frac{\rho}{8}$, by (7) we have

(10)
$$\sum_{j=0}^{n-3} ((z_j - z_0) \cdot a)^2 \geq \frac{\rho^2}{8}|a|^2, \quad a \in L,$$

where L is the linear subspace spanned by $z_j - z_0$, $j = 1, \ldots, n-3$. Notice that automatically L satisfies

(11)
$$\|L - \{0\} \times \mathbb{R}^{n-3}\| \leq C\delta$$

by virtue of (2), (10) and the fact that $z_0, \ldots, z_{n-3} \in S_+ \cap B_{2\rho}(w_0)$.

Similarly, for arbitrary given $w \in (w_0 + (\{0\} \times \mathbb{R}^{n-3})) \cap B_\rho(w_0)$, and for any set $\zeta_0^0, \ldots, \zeta_{n-3}^0 \in B_{\theta\rho}(w) \cap (w_0 + \{0\} \times \mathbb{R}^{n-3})$ with

(12)
$$\sum_{j=0}^{n-3}((\zeta_j^0 - \zeta_0^0) \cdot a)^2 \geq \frac{\theta^2 \rho^2}{2}|a|^2, \quad a \in \{0\} \times \mathbb{R}^{n-3},$$

we can again use the general principle (9). This time we in fact use (9) with the choices $U = B_{\theta\rho}(w)$ and $U = B_\rho(w_0) \setminus \{(x,y) : |x - \xi_{w_0}| \leq \frac{\rho}{8}\}$, in each case taking $V = B_{\theta\rho/4}(\zeta_j^0) \cap S_+$. Then, keeping in mind (5), (6), the fact that $\theta \geq 8\delta^{1/2}$ and the remark immediately following (9), we deduce that we can select $\zeta_j \in B_{\theta\rho/8}(\zeta_j^0) \cap S_+$ such that for each $j = 0, \ldots, n-3$

(13)
$$\int_{B_{\theta\rho}(w)} \frac{|R_{\zeta_j} u_{R_{\zeta_j}}|^2}{R_{\zeta_j}^n} \leq C(\theta\rho)^{-(n-3)} \int_{B_{\theta\rho/4}(w)} \psi$$

$$\int_{B_\rho(w_0) \setminus \{(x,y):|x-\xi_{w_0}|\leq\rho/8\}} \frac{|R_{\zeta_j} u_{R_{\zeta_j}}|^2}{R_{\zeta_j}^n} \leq C(\theta\rho)^{-(n-3)} \int_{B_\rho(w_0) \setminus \{(x,y):|x-\xi_{w_0}|\leq\rho/8\}} \psi,$$

where $C = C(n, N, \beta)$. (We emphasize that the choice of ζ_j depends on w, but C only depends on n, N, β.) Since $|\zeta_j - \zeta_j^0| \leq \frac{\theta\rho}{8}$, from (11) and (12) we also deduce that

(14) $$\sum_{j=0}^{n-3}((\zeta_j - \zeta_0) \cdot a)^2 \geq \frac{\theta^2 \rho^2}{8}|a|^2, \quad a \in L.$$

Now in view of (10) and (14) we can apply Lemma 1 of Section 4.5 in order to conclude

(15) $$\sum_{j=0}^{n-3}|R_{z_j}u_{R_{z_j}}|^2 \leq C\theta^{-2}\sum_{j=0}^{n-3}|R_{\zeta_j}u_{R_{\zeta_j}}|^2 + C\theta^{-2}\operatorname{dist}^2(\zeta_j, z_0 + L)|Du|^2$$

on $B_{\theta\rho}(w)$, and hence

(16) $$\int_{B_{\theta\rho}(w)} \sum_{j=0}^{n-3}|R_{z_j}u_{R_{z_j}}|^2 \leq$$

$$\leq C\theta^{-2}\int_{B_{\theta\rho}(w)}\sum_{j=0}^{n-3}|R_{\zeta_j}u_{R_{\zeta_j}}|^2 + C\theta^{-2}\int_{B_{\theta\rho}(w)}|Du|^2 \operatorname{dist}^2(\zeta_j, z_0 + L)$$

$$\leq C\theta^{-2}(\theta\rho)^n \int_{B_{\theta\rho}(w)}\sum_{j=0}^{n-3}\frac{|R_{\zeta_j}u_{R_{\zeta_j}}|^2}{R_{\zeta_j}^n} + C\theta^{-2}(\theta\rho)^{n-2} \operatorname{dist}^2(\zeta_j, z_0 + L)$$

$$\leq C\theta\rho^3 \int_{B_{\theta\rho}(w)} \psi + C\theta^{n-4}\rho^{n-2} \operatorname{dist}^2(\zeta_j, z_0 + L)$$

by the first inequality in (13). Now by Lemma 1 and Lemma 2 of Section 4.5 together with (8) and the second inequality in (13) we have

(17) $$\operatorname{dist}^2(\zeta_j, z_0 + L) \leq C\rho^2 \int_{B_\rho(w_0)\setminus\{(x,y):|x-\xi_{w_0}|\leq \rho/8\}} \sum_{j=0}^{n-3}\left(\frac{|R_{z_j}u_{R_{z_j}}|^2}{R_{z_j}^n} + \frac{|R_{\zeta_j}u_{R_{\zeta_j}}|^2}{R_{\zeta_j}^n}\right)$$

$$\leq C\rho^2(\theta\rho)^{-(n-3)} \int_{B_\rho(w_0)\setminus\{(x,y):|x-\xi_{w_0}|\leq \theta\rho/8\}} \psi.$$

Thus by combining (16) and (17) we conclude

(18) $$\int_{B_{\theta\rho}(w)}\sum_{j=0}^{n-3}|R_{z_j}u_{R_{z_j}}|^2 \leq C\theta\rho^3 \int_{B_{\theta\rho}(w)} \psi + C\theta^{-1}\rho^3 \int_{B_\rho(w_0)\setminus\{(x,y):|x-\xi_{w_0}|<\theta\rho/8\}} \psi$$

for each $w \in (w_0 + (\{0\} \times \mathbb{R}^{n-3})) \cap B_\rho(w_0)$.

The presence of the factor θ in the first term here is crucial, as we shall see below.

Now we are going to use the main L^2-estimate from Theorem 1 of Section 4.5 with

4.6. The deviation function ψ

$\frac{\rho}{2}$ in place of ρ. Thus (since $B_{\rho/16}(z_0) \supset B_{\rho/32}(w_0)$ and $B_{\rho/2}(z_0) \subset B_\rho(w_0)$)

(19)
$$\int_{B_{\rho/32}(w_0)} \frac{|R_z u_{R_z}|^2}{R_z^n} \leq C\rho^{-2} s(z)^{2-n} \int_{B_\rho(w_0)} \sum_{j=0}^{n-3} |R_{z_j} u_{R_{z_j}}|^2 + C\left(\frac{\rho^{n-2}}{s(z)^{n-2}}\right)^{1-1/(2-\alpha)} \times$$
$$\times \left(\int_{B_\rho(w_0) \setminus \{(x,y):|x-\xi_{w_0}| \leq \rho/16\}} \left(\frac{1}{\rho^2 s(z)^{n-2}} \sum_{j=0}^{n-3} |R_{z_j} u_{R_{z_j}}|^2 + \frac{|R_z u_{R_z}|^2}{R_z^n}\right)\right)^{1/(2-\alpha)}$$

for each $z \in S_+$.

Now we want to integrate this with respect to the measure μ^+. First notice that, using the notation $\mu^+(A) = \mu^+(A \cap S_+)$,

(20)
$$\int_{S_+} \frac{1}{s(z)^{n-2}} d\mu^+ \leq \frac{\mu^+(B_\rho(w_0))}{\rho^{n-2}} + \sum_{j=1}^{\infty} \frac{\mu^+(B_{(j+1)\rho}(w_0)) - \mu^+(B_{j\rho}(w_0))}{(j\rho)^{n-2}}$$
$$\leq C\rho^{-1} + C\rho^{2-n} \sum_{j=1}^{\infty} \mu^+(B_{(j+1)\rho}(w_0))(j^{2-n} - (j+1)^{2-n})$$
$$\leq C\rho^{-1} + C\rho^{-1} \sum_{j=1}^{\infty} j^{-2} \leq C\rho^{-1},$$

where we used summation by parts and the fact that $\mu^+(B_{(j+1)\rho}(w_0)) \leq Cj^{n-3}\rho^{n-3}$ by virtue of Lemma 1 of Section 4.3. Thus integrating in (19) and using the Hölder inequality and (20) we deduce that

(21)
$$\int_{B_{\rho/32}(w_0)} \psi \leq C\rho^{-3} \int_{B_\rho(w_0)} \sum_{j=0}^{n-3} |R_{z_j} u_{R_{z_j}}|^2 +$$
$$+ C(\rho^{n-3})^{1-1/(2-\alpha)} \left(\int_{B_\rho(w_0) \setminus \{(x,y):|x-\xi_{w_0}| \leq \rho/16\}} (\rho^{-3} \sum_{j=0}^{n-3} |R_{z_j} u_{R_{z_j}}|^2 + \psi)\right)^{1/(2-\alpha)}.$$

Now we select points $w_1, \ldots, w_Q$ (with $Q = Q(n, \theta)$) in $B_{3\rho/4}(w_0) \cap (w_0 + (\{0\} \times \mathbb{R}^{n-3}))$ such that $\{B_{\theta\rho/16}(w_j)\}$ are pairwise disjoint and such that $\{B_{\theta\rho/4}(w_j)\}$ cover the $\frac{\theta\rho}{8}$-neighbourhood of $B_{3\rho/4}(w_0) \cap (w_0 + (\{0\} \times \mathbb{R}^{n-3}))$. Then (21) implies

(22)
$$\int_{B_{\rho/32}(w_0)} \psi \leq C\rho^{-3} \sum_{i=1}^{Q} \int_{B_{\theta\rho/4}(w_i)} \sum_{j=0}^{n-3} |R_{z_j} u_{R_{z_j}}|^2 +$$
$$+ C\rho^{-3} \int_{B_\rho(w_0) \setminus \{(x,y):|x-\xi_{w_0}| \leq \theta\rho/8\}} \sum_{j=0}^{n-3} |R_{z_j} u_{R_{z_j}}|^2$$
$$+ C(\rho^{n-3})^{1-1/(2-\alpha)} \left(\int_{B_\rho(w_0) \setminus \{(x,y):|x-\xi_{w_0}| \leq \theta\rho/8\}} \rho^{-3} \sum_{j=0}^{n-3} |R_{z_j} u_{R_{z_j}}|^2 + \psi\right)^{1/(2-\alpha)}.$$

Now we use (18) with w_i in place of w, thus estimating the terms $\int_{B_{\theta\rho/4}(w_i)} \sum_{j=0}^{n-3} |R_{z_j} u_{R_{z_j}}|^2$ on the right. At the same time we can use (8) in the remaining terms on the right.

Thus we obtain from (22) that

$$(23) \quad \int_{B_{\rho/32}(w_0)} \psi \leq C\theta \sum_{j=1}^{Q} \int_{B_{\theta\rho/4}(w_j)} \psi + C_1 \int_{B_\rho(w_0)\setminus\{(x,y):|x-\xi_{w_0}|\leq \theta\rho/8\}} \psi +$$
$$+ C_1(\rho^{n-3})^{1-1/(2-\alpha)} \left(\int_{B_\rho(w_0)\setminus\{(x,y):|x-\xi_{w_0}|\leq \theta\rho/8\}} \psi \right)^{1/(2-\alpha)},$$

where $C = C(n,N,\beta)$ is independent of θ and $C_1 = C_1(\theta,n,N,\beta)$. Now notice that $B_\rho(w_0) \cap S_+ \subset \{(x,y) : |x - \xi_{z_0}| < 6\delta\rho\}$ by (4), and $\rho^{n-3} \leq C\mu^+(B_\rho(w_0))$ by Lemma 1 of Section 4.3, where we continue to use the convention $\mu^+(A) = \mu^+(A \cap S_+)$. Hence, assuming $6\delta \leq \frac{\theta}{16}$, we see that (23) implies

$$(24) \quad \int_{B_{\rho/32}(w_0)} \psi \leq C\theta \int_{B_\rho(w_0)} \psi + C_1 \int_{B_\rho(w_0)\setminus S^+_{\theta\rho/16}} \psi +$$
$$+ C_1(\mu^+(B_\rho(w_0)))^{1-1/(2-\alpha)} \left(\int_{B_\rho(w_0)\setminus S^+_{\theta\rho/16}} \psi \right)^{1/(2-\alpha)},$$

where $S^+_\sigma = \{(x,y) : \text{dist}((x,y), S_+) < \sigma\}$. Now notice that this was all valid starting with an arbitrary $w_0 \in T_\rho \cap S_+$. Now choose a maximal pairwise-disjoint collection $\{B_{\rho/128}(p_k)\}_{k=1,\ldots,P}$ with $p_k \in S_+ \cap T^+_{\rho/4}$. Then $\cup B_{\rho/32}(p_k)$ covers all of the $\frac{\rho}{64}$ neighbourhood of $S_+ \cap T^+_{\rho/4}$. Notice that by Remark (iv)(2)(a) of Section 4.3 we have also that $\cup B_\rho(p_k)$ is contained in $T^+_{2\rho}$. Since any point of $T^+_{2\rho}$ lies in at most $C(n)$ of the balls $B_\rho(p_k)$, we then have, by replacing w_0 by p_k in (24) and summing over k,

$$\int_{T^+_{\theta\rho}} \psi \leq C\theta \int_{T^+_{2\rho}} \psi + C \int_{T^+_{2\rho}\setminus S^+_{\theta\rho/16}} \psi + C_1(\mu^+(T^+_{2\rho}))^{1-1/(2-\alpha)} \left(\int_{T^+_{2\rho}\setminus S^+_{\theta\rho/16}} \psi \right)^{1/(2-\alpha)}$$
$$\leq C\theta \int_{T^+_{\theta\rho}} \psi + C\theta \int_{T^+_{2\rho}\setminus T^+_{\theta\rho}} \psi + C \int_{T^+_{2\rho}\setminus T^+_{\theta\rho/16}} \psi +$$
$$+ C_1(\mu^+(T^+_{2\rho}))^{1-1/(2-\alpha)} \left(\int_{T^+_{2\rho}\setminus S^+_{\theta\rho/16}} \psi \right)^{1/(2-\alpha)}$$
$$\leq C\theta \int_{T^+_{\theta\rho}} \psi + C \int_{T^+_{2\rho}\setminus T^+_{\theta\rho/16}} \psi + C_1(\mu^+(T^+_{2\rho}))^{1-1/(2-\alpha)} \left(\int_{T^+_{2\rho}\setminus S^+_{\theta\rho/16}} \psi \right)^{1/(2-\alpha)},$$

so that since $\mu^+(S_+) = 1$ and $S^+_\sigma \supset T^+_\sigma$ we get finally that

$$\int_{T^+_{\theta\rho/16}} \psi \leq C \left(\int_{T^+_{2\rho}\setminus T^+_{\theta\rho/16}} \psi \right)^{1/(2-\alpha)},$$

provided $\theta = \theta(n, N, \beta) \in (0, \frac{1}{32})$ is chosen so that $C\theta \leq \frac{1}{2}$ and provided $8\delta^{1/2} \leq \theta$ (i.e. $\delta \leq \delta_0$, with $\delta_0 = \frac{\theta^2}{64}$). By a change of notation (taking $\frac{\theta}{16}$ to 2θ) and replacing ρ by $\frac{\rho}{2}$, we then have the required inequality. $\square$

4.7 Proof of Theorems 1, 2 of Section 4.1

Let $\beta > 0$, let $\theta_0 \in \{\Theta_\varphi(0) : \varphi$ is an HCM with $\Theta_\varphi(0) \leq \beta\}$ be arbitrary, and suppose

(i) $\qquad w_0 \in \text{sing } u$ with $\Theta_u(w_0) = \theta_0$.

Recall that, by the monotonicity identity, for each $\varepsilon \in (0, 1)$ there exists $\sigma_0 = \sigma_0(\varepsilon, u, w_0) > 0$ such that

(ii) $\qquad \Theta_u(w_0) \leq \sigma^{2-n} \int_{B_\sigma(w_0)} |Du|^2 \leq \Theta_u(w_0) + \varepsilon, \quad \sigma \in (0, \sigma_0].$

Also, by monotonicity (see (i) of Section 2.4) we have the identity

(iii) $\qquad 2 \int_{B_\rho(z)} \frac{|R_z u_{R_z}|^2}{R_z^n} = \sigma^{2-n} \int_{B_\sigma(z)} |Du|^2 - \Theta_u(z)$

for each z, ρ such that $\overline{B}_\rho(z) \subset \Omega$, and, since $B_\sigma(z) \subset B_{(1+\varepsilon)\sigma}(w_0)$ for any $z \in B_{\varepsilon\sigma}(w_0)$, we deduce from (ii) that

$$2 \int_{B_\sigma(z)} \frac{|R_z u_{R_z}|^2}{R_z^n} = \sigma^{2-n} \int_{B_\sigma(z)} |Du|^2 - \Theta_u(z)$$
$$\leq (1+\varepsilon)^{n-2}((1+\varepsilon)\sigma)^{2-n} \int_{B_{(1+\varepsilon)\sigma}(w_0)} |Du|^2 - \Theta_u(z)$$
$$\leq C(n)(1+\beta)\varepsilon, \quad z \in B_{\varepsilon\sigma}(w_0), \quad \sigma \leq \frac{\sigma_0}{2},$$

provided $\Theta_u(z) \geq \theta_0$ and provided $\sigma_0 = \sigma_0(u, w_0, \varepsilon) > 0$ is sufficiently small. Let

$$S_+ = \{z \in \overline{B}_{\varepsilon\sigma_0/2}(w_0) : \Theta_u(z) \geq \theta_0\},$$

take $w_1 \in S_+ \cap \overline{B}_{\varepsilon\sigma_0/4}(w_0)$, $\sigma_1 \in (0, \frac{\varepsilon\sigma_0}{4}]$ and define

(iv) $\qquad \tilde{u} = u_{w_1, \sigma_1},$

where $u_{w_1, \sigma_1}(x, y) \equiv u(w_1 + \sigma_1(x, y))$. Then the above inequality gives

(v) $\qquad 2 \int_{B_\rho(z)} \frac{|R_z \tilde{u}_{R_z}|^2}{R_z^n} = \rho^{2-n} \int_{B_\rho(z)} |D\tilde{u}|^2 - \Theta_{\tilde{u}}(z) \leq C\varepsilon, \quad z \in S_+(w_1, \sigma_1), \quad \rho \in (0, \frac{1}{4}],$

where $C = C(n, \beta)$ and

(vi) $\quad 0 \in S_+(w_1, \sigma_1) \equiv \{z \in \overline{B}_1(0) : \Theta_{\tilde{u}}(z) \geq \theta_0\} = \overline{B}_1(0) \cap \eta_{w_1, \sigma_1} S_+.$

Notice that $S_+(w_1, \sigma_1)$ corresponds exactly to the S_+ of Section 4.5 and 4.6 with $\tilde{u}$ in place of u. Also, recall that by Lemma 2 of Section 3.9, we can, and we shall, assume that $\sigma_0 = \sigma_0(u, w_0, \varepsilon)$ is chosen small enough so that S_+ has the ε-approximation property of (i) Section 4.2 and hence $S_+(w_1, \sigma_1)$ does also. Thus (Cf. (iv) of Section 4.5)

(vii) $\quad S_+(w_1, \sigma_1) \cap B_\sigma(z) \subset$ the $(\varepsilon\sigma)$-neighbourhood of $L_{z,\sigma}$,

for each $z \in S_+(w_1, \sigma_1)$ and each $\sigma \in (0, 1]$, where $L_{z,\sigma}$ is an $(n-3)$-dimensional affine space containing z. We fix these affine spaces in the sequel. Without loss of generality we assume

(viii) $\quad L_{0,1} = \{0\} \times \mathbb{R}^{n-3}.$

We emphasize that (v) and (vii) hold automatically if $\sigma_0 = \sigma_0(\varepsilon, u, w_0)$ is chosen sufficiently small. We henceforth assume $\sigma_0(\varepsilon, u, w_0)$ has been so chosen, and we continue to take $\tilde{u}$ as in (iv). Notice also that, by (iv) (choosing new ε if necessary), (i), (iii), (v) of Section 4.5 also hold with $S_+ = S_+(w_1, \sigma_1)$. Thus we can apply the results of Section 4.5 and 4.6 with $\tilde{u}$ in place of u, and with $S_+ = S_+(w_1, \sigma_1)$, $\theta_0 = \Theta_u(w_0)$.

Before we begin, we need to establish the following lemma, which is a simple inequality for real numbers:

Lemma 1 *If $0 < a < b \leq 1$, $\alpha \in (0, 1)$, $\beta > 0$ and $a^{2-\alpha} \leq \beta(b-a)$, then*

$$a^{-1+\alpha/2} - b^{-1+\alpha/2} \geq Ca^{-\alpha/2}, \quad C = C(\beta, \alpha) > 0.$$

Proof: In case $\frac{b}{a} > 2$ we have trivially that

$$a^{-1+\alpha/2} - b^{-1+\alpha/2} \geq Ca^{-1+\alpha/2} \geq Ca^{-\alpha/2},$$

so the required inequality holds in this case. In case $\frac{b}{a} \leq 2$ we have

$$\begin{aligned}
a^{-1+\alpha/2} - b^{-1+\alpha/2} &= (1-\alpha/2)c^{-2-\alpha/2}(b-a) \quad \text{for some } c \in (a, b) \\
&\geq \frac{1-\alpha/2}{4} a^{-\alpha/2} \frac{b-a}{a^{2-\alpha}} \quad \text{since } a \geq \frac{b}{2} \\
&\geq \frac{\beta(1-\alpha/2)}{4} a^{-\alpha/2} \quad \text{since } a^{2-\alpha} \leq \beta(b-a),
\end{aligned}$$

so again the required inequality is satisfied, and the lemma is proved. $\square$

We now give the proof of Theorem 1. We shall only need the real-analyticity hypothesis in checking the Łojasiewicz inequality for the energy functional as in Section 3.14. Since we already checked in Section 3.14 that this Łojasiewicz inequality

4.7. Proof of Theorems 1, 2 of Section 4.1

holds automatically (with any exponent $\alpha \in (0,1]$) when the integrability condition of Theorem 2 holds, we thus see that Theorem 2 will follow by exactly the same argument used to prove Theorem 1.

Proof of Theorem 1:

Let T_ρ^+, μ^+ (corresponding to given δ with $\varepsilon < \frac{\delta}{8}$, and with $\tilde{u}$ as in (iv) in place of u) be as in Section 4.6. $\delta \leq \delta_0(n, N, \beta)$ and $\varepsilon < \frac{\delta}{8}$ will be chosen later.

Now, with $\tilde{u}$ as in (iv), by virtue of (iii), (v), we can apply all the results of Section 4.6 to $\tilde{u}$, and hence

$$\text{(1)} \qquad \int_{T_{\theta\rho}^+} \psi \leq C \left(\int_{T_\rho^+ \setminus T_{\theta\rho}^+} \psi \right)^{1/(2-\alpha)}$$

with ψ the deviation function of Section 4.6 with $\tilde{u}$ in place of u, where $\theta = \theta(n, N, \beta) > 0$, and where $\alpha = \alpha(n, N, \beta) \in (0,1)$.

In view of Lemma 1 we can use (1) to give

$$\text{(2)} \qquad \left(\int_{T_{\theta\rho}^+} \psi \right)^{-1+\alpha/2} - \left(\int_{T_\rho^+} \psi \right)^{-1+\alpha/2} \geq C I_0^{-\alpha/2},$$

where $I_0 = \int_{T_{1/4}^+} \psi$. Then starting with $\rho = \frac{1}{4}$ we can iterate the inequality (2) in order to give

$$\left(\int_{T_{\theta^j/4}^+} \psi \right)^{-1+\alpha/2} \geq Cj I_0^{-\alpha/2}, \quad j = 1, 2, \dots,$$

and hence

$$\text{(3)} \qquad \int_{T_{\theta^j/4}^+} \psi \leq C j^{-1-2\gamma} I_0^{2\gamma}, \quad j = 1, 2, \dots,$$

where $2\gamma = \frac{\alpha}{2-\alpha} > 0$. Notice that since $(j+1)^{1+\gamma} - j^{1+\gamma} \geq Cj^\gamma$, this implies

$$\sum_{j=0}^\infty ((j+1)^{1+\gamma} - j^{1+\gamma}) \int_{T_{\theta^j/4}^+} \psi \leq CI_0^{2\gamma} \sum_{j=1}^\infty j^{-1-\gamma} \leq CI_0^{2\gamma},$$

and using summation by parts we thus have

$$\sum_{j=1}^\infty j^{1+\gamma} \int_{T_{\theta^{j-1}/4}^+ \setminus T_{\theta^j/4}^+} \psi \leq CI_0^{2\gamma}.$$

Thus we get

$$\text{(4)} \qquad \int_{T_{1/4}^+} |\log d|^{1+\gamma} \psi \leq CI_0^{2\gamma},$$

where d is defined on $T_{1/4}^+$ by

(5) $$d(x,y) = \begin{cases} 2^{-k} & \text{if } (x,y) \in T_{2^{-k}}^+ \setminus T_{2^{-k-1}}^+, \ k \geq 2 \\ 0 & \text{if } (x,y) \in T_0^+. \end{cases}$$

Now for $z \in S_+ \cap T_{1/4}^+$ and $(x,y) \in T_{1/4}^+$ we claim that

(6) $$d(x,y) \leq 4R_z(x,y), \quad (x,y) \in T_{1/4}^+ \setminus B_{d(z)/2}(z)$$

where $R_z(x,y) = |(x,y) - z|$. Here we include $z \in T_0^+$, in which case $d(z) = 0$ so (6) says $d(x,y) \leq 4R_z(x,y), \ \forall (x,y) \in T_{1/4}^+$. To prove this we can of course assume $d(x,y) > 0$, so take any $w = (x,y) \in T_{2^{-k}}^+ \setminus T_{2^{-k-1}}^+$ for some $k \geq 2$, and consider cases as follows:

Case (a): $z \in T_{2^{-q}}^+$ for some $q \geq k+2$. (If $z \in T_0^+$, then this case will be applicable $\forall q \geq k+2$.) Then by Remark (iv)(2)(d) of Section 4.3 we have $|w-z| \geq 2^{-k-2} = \frac{2^{-k}}{4} = \frac{d(w)}{4}$.

Case (b): $z \in T_{2^{-q}}^+ \setminus T_{2^{-q-1}}^+$ with $q \leq k+1$. In this case, if we assume that $w \notin B_{d(z)/2}(z)$ then (keeping in mind that $z \in S_+$ and $d(z) = 2^{-q}$ in case $z \in T_{2^{-q}}^+ \setminus T_{2^{-q-1}}^+$), we have $|w-z| \geq 2^{-q-1} \geq 2^{-k-2} = \frac{d(w)}{4}$.

Thus (6) is always satisfied as claimed. Now the inequality (4) says that

(7) $$\int_{T_{1/4}^+} |\log d|^{1+\gamma} \int_{S_+} \frac{|R_z \tilde{u}_{R_z}|^2}{R_z^n} d\mu^+(z) \, dxdy \leq CI_0^{2\gamma},$$

so that by interchanging the order of integration we deduce that

(8) $$\int_{T_{1/4}^+} |\log d|^{1+\gamma} \frac{|R_z \tilde{u}_{R_z}|^2}{R_z^n} dxdy \leq I_0^\gamma,$$

for all $z \in S_+$ with the exception of a set of μ^+-measure $\leq CI_0^\gamma$. (We must keep in mind here that there will in general be lots of points $z \in S_+$ which are not in the support of μ^+, and these have μ^+-measure zero, so in particular (8) need not hold for them.)

In view of (6), (8) implies

(9) $$\int_{T_{1/4}^+ \setminus B_{d(z)/2}(z)} |\log R_z|^{1+\gamma} \frac{|R_z \tilde{u}_{R_z}|^2}{R_z^n} dxdy \leq I_0^\gamma,$$

for all $z \in S_+$ with the exception of a set of μ^+-measure $\leq CI_0^\gamma$.

Next note that according to Lemma 1 of Section 4.3 we have a countable set $\mathcal{S} = \{z_{j,k} : j = 1, \ldots, Q_k, \ k \geq 2\} \subset S_+ \cap T_{1/4}^+$ such that

(10) $\quad z_{j,k} \in T_{2^{-k}}^+ \setminus T_{2^{-k-1}}^+$, so $d(z_{j,k}) = 2^{-k}, \quad j = 1, \ldots, Q_k, \ k \geq 2,$

4.7. Proof of Theorems 1, 2 of Section 4.1

(11) $$\mu = C(\delta^{(n-3)/2} \sum_{k=2}^{\infty} 2^{-(n-3)k} \sum_{j=1}^{Q_k} [\![z_{k,j}]\!] + \mathcal{H}^{n-3} \llcorner T_0^+), \quad C = C(n),$$

and

(12) $$S \cap T_{2^{-k}}^+ \setminus T_{2^{-k-1}}^+ \subset \bigcup_{\ell=\max(k-2,2)}^{k+1} \bigcup_{j=1}^{Q_\ell} B_{\delta^{1/2} 2^{-k}}(z_{\ell,j}) \quad \forall k \geq 2.$$

Now let $\mathcal{E}_0 \subset \mathcal{S}$ be the collection of all $z_{j,k} \in \mathcal{S}$ such that

(13) $$\int_{T_{1/4}^+ \setminus B_{d(z_{j,k})/2}(z_{j,k})} |\log R_{z_{j,k}}|^{1+\gamma} \frac{|R_{z_{j,k}} \tilde{u}_{R_{z_{j,k}}}|^2}{R_{z_{j,k}}^n} \, dx \, dy \geq I_0^\gamma,$$

and let $\mathcal{E}_1 \subset T_0^+$ be the collection of all $z \in T_0^+$ such that

(14) $$\int_{T_{1/4}^+} |\log R_z|^{1+\gamma} \frac{|R_z \tilde{u}_{R_z}|^2}{R_z^n} \, dx \, dy \geq I_0^\gamma.$$

Since $\mu^+(\mathcal{E}_0 \cup \mathcal{E}_1) \leq C I_0^\gamma$ by (9), we thus have by (11) that

(15) $$\sum_{w \in \mathcal{E}_0} d(w)^{n-3} + \mathcal{H}^{n-3}(\mathcal{E}_1) \leq C I_0^\gamma, \quad C = C(n, N, \delta).$$

Now take any $z \in T_{1/4}^+ \cap S_+ \setminus T_0^+$. We have by (12) that $z \in B_{d(z_{j,k})/4}(z_{j,k})$ for some $z_{j,k} \in \mathcal{S}$; if this $z_{j,k} \notin \mathcal{E}_0$ then by (9)

(16) $$\int_{T_{1/4}^+ \setminus B_{d(z_{j,k})/2}(z_{j,k})} |\log R_{z_{j,k}}|^{1+\gamma} \frac{|R_{z_{j,k}} \tilde{u}_{R_{z_{j,k}}}|^2}{R_{z_{j,k}}^n} \, dx \, dy \leq I_0^\gamma.$$

Regardless of whether $z_{j,k} \in \mathcal{E}_0$ or not, we have by (10) and Remark (iv)(2)(d) of Section 4.3 (with $k+2$, $k+1$ in place of ℓ, k) that $z \in B_{d(z_{j,k})/4}(z_{j,k}) \subset \mathbb{R}^n \setminus T_{2^{-k-2}}^+$ and hence that

(17) $$d(z) \geq 2^{-k-1} = \tfrac{1}{2} d(z_{j,k}).$$

Thus by (16), (17), for any $z \in S_+ \cap T_{1/4}^+ \setminus T_0^+$,

(18) either $z \in S_+ \cap (\cup_{z_{j,k} \in \mathcal{E}_0} B_{d(z_{j,k})/4}(z_{j,k}))$
or $\exists \tilde{z}$ ($=$ some $z_{j,k} \in S_+ \cap T_{1/4}^+ \setminus (T_0^+ \cup \mathcal{E}_0)$) with $d(z) \geq \tfrac{1}{2} d(\tilde{z})$,
$z \in B_{d(\tilde{z})/4}(\tilde{z})$, and $\int_{T_{1/4}^+ \setminus B_{d(\tilde{z})/2}(\tilde{z})} |\log R_{\tilde{z}}|^{1+\gamma} \frac{|R_{\tilde{z}} \tilde{u}_{R_{\tilde{z}}}|^2}{R_{\tilde{z}}^n} \, dx \, dy \leq I_0^\gamma.$

On the other hand if $z \in T_0^+ \setminus \mathcal{E}_1$ then by definition $d(z) = 0$ and (9) gives

(19) $$\int_{T_{1/4}^+} |\log R_z|^{1+\gamma} \frac{|R_z \tilde{u}_{R_z}|^2}{R_z^n} \, dx \, dy \leq I_0^\gamma.$$

Now we want to check that we have all the hypotheses needed to apply the rectifiability lemma of Section 4.2 (in case $\rho = 1$, $m = n - 3$, and $S = S_+$).

To check these hypotheses, we first assume

(20) no ball $B_\rho(z)$ with $z \in \overline{B}_{5/8}(0) \cap S_+$ and $\rho \in [\frac{1}{8}, \frac{1}{4}]$ has a δ-gap.

Note that then by (20), (vii), (viii), and the fact that $\varepsilon < \frac{\delta}{8}$ we have

(21) $\qquad B_{5/8}(0) \cap \{(x, y) : |x| \leq \frac{1}{8}\} \subset T_{1/4}^+$.

In this case (19) implies

(22) $\qquad \int_{B_{5/8}(0) \setminus B_{d(z)/2}(z)} |\log R_z|^{1+\gamma} \frac{|R_z \tilde{u}_{R_z}|^2}{R_z^n} \, dx dy \leq I_0^\gamma,$

for any $z \in T_0^+ \setminus \mathcal{E}_1$, and (18) implies that for any point z in the subset $(S_+ \cap \overline{B}_{1/2}(0) \setminus T_0^+) \setminus (\cup_{z_{j,k} \in \mathcal{E}_0} B_{d(z_{j,k})/4}(z_{j,k}))$ there is always a point $\tilde{z} \in S_+ \cap T_{1/4}^+ \setminus T_0^+$ such that

(23)
$$\int_{B_{5/8}(0) \setminus B_{d(\tilde{z})/2}(\tilde{z})} |\log R_{\tilde{z}}|^{1+\gamma} \frac{|R_{\tilde{z}} \tilde{u}_{R_{\tilde{z}}}|^2}{R_{\tilde{z}}^n} \, dx dy \leq I_0^\gamma, \quad z \in B_{d(\tilde{z})/4}(\tilde{z}), \quad d(z) \geq \tfrac{1}{2} d(\tilde{z}).$$

Our aim now is to show that the hypotheses of the rectifiability lemma of Section 4.2 are satisfied.

First take an arbitrary point $z \in (S_+ \cap \overline{B}_{1/2}(0) \setminus T_0^+) \setminus (\cup_{z_{j,k} \in \mathcal{E}_0} B_{d(z_{j,k})/4}(z_{j,k}))$ and let $\tilde{z}$ be as in (23). Using the notation that $\hat{u}(\sigma)$ denotes the $L^2(S^{n-1})$ function given by $\hat{u}(s)(\omega) = \hat{u}(\tilde{z} + s\omega)$, $\omega \in S^{n-1}$, we then have by direct integration, the Cauchy-Schwarz inequality, and (23) that

(24) $\|\hat{u}(\sigma) - \hat{u}(\tau)\|_{L^2(S^{n-1})} \leq$

$$\leq \int_\sigma^\tau \left\| \frac{\partial \hat{u}(s)}{\partial s} \right\|_{L^2(S^{n-1})} ds$$

$$\leq \left(\int_\sigma^\tau |\log s|^{1+\gamma} s \left\| \frac{\partial \hat{u}(s)}{\partial s} \right\|_{L^2(S^{n-1})}^2 ds \right)^{1/2} \left(\int_\sigma^\tau s^{-1} |\log s|^{-1-\gamma} ds \right)^{1/2}$$

$$\leq \gamma^{-1/2} \left(\int_{B_{5/8}(0) \setminus B_{d(\tilde{z})/2}(\tilde{z})} |\log R_{\tilde{z}}|^{1+\gamma} \frac{|R_{\tilde{z}} \tilde{u}_{R_{\tilde{z}}}|^2}{R_{\tilde{z}}^n} \right)^{1/2} |\log \tau|^{-\gamma/2}$$

$$\leq C I_0^{\gamma/2} |\log \tau|^{-\gamma/2}$$

for any $\frac{3d(z)}{4} < \sigma < \tau \leq \frac{1}{4}$. Notice also that by applying Lemma 2 of Section 4.5 (keeping in mind (vii)), we have that $\|\tilde{u} - \varphi\|_{L^2(B_{1/4}(z))}^2 \leq C\varepsilon$ for some HCM φ with $S(\varphi) = \{0\} \times \mathbb{R}^{n-3}$, and also from hypothesis (v) we know that $I_0 \leq C\varepsilon$. So if $\zeta \leq \delta^2$ is given and if ε is small enough (depending on $n, N, \beta, \delta, \zeta$) we then deduce from (24), Corollary 1 of Section 4.4 and (vii) that

(25) $\qquad \operatorname{sing} \tilde{u} \cap B_\rho(\tilde{z}) \subset$ the $(\zeta \rho)$-neighbourhood of $\tilde{z} + \{0\} \times \mathbb{R}^{n-3}$, $\quad \forall \rho \in (\tfrac{d(z)}{4}, \tfrac{1}{2}]$.

4.7. Proof of Theorems 1, 2 of Section 4.1

Now, still assuming $z \in (S_+ \setminus T_0^+) \cap \overline{B}_{1/2}(0) \setminus (\cup_{z_{j,k} \in \mathcal{E}_0} B_{d(z_{j,k})/4}(z_{j,k}))$, (25) guarantees in particular that $S_+ \cap B_\rho(z) \subset$ the $(\delta\rho)$-neighbourhood of $z + \{0\} \times \mathbb{R}^{n-3}$ for every $\rho \in [\frac{d(z)}{2}, \frac{1}{2}]$, so (using the definition (iii) of Section 4.3) $z \in T^+_{d(z)/2}$ unless one of the balls $B_\rho(z)$ has a δ-gap. But of course $z \in T^+_{d(z)/2}$ contradicts the definition of $d(z)$ for $d(z) > 0$, so we conclude finally (keeping in mind (20))

(26) $\forall z \in (S_+ \setminus T_0^+) \cap \overline{B}_{1/2}(0) \setminus (\bigcup_{z_{j,k} \in \mathcal{E}_0} B_{d(z_{j,k})/4}(z_{j,k}))$,

$$\exists \sigma_z \in [\tfrac{d(z)}{2}, \tfrac{1}{8}] \text{ such that } S_+ \text{ has a } \delta\text{-gap in } B_{\sigma_z}(z).$$

Next notice that since T_0^+ is a subset of the graph of a Lipschitz function over $\{0\} \times \mathbb{R}^{n-3}$ with Lipschitz constant $\leq C\delta$, in view of (15) we can select a $\{B_{\sigma_k}(z_k)\}$ such that

(27) $\qquad \sigma_k \in (0, \tfrac{1}{8}), \quad \mathcal{E}_1 \subset \cup_k B_{\sigma_k}(z_k), \quad \sum_k \sigma_k^{n-3} \leq C I_0^\gamma.$

For $z \in S_+ \cap T_0^+ \setminus \cup_k B_{\sigma_k}(z_k)$ we have, by the same argument that we used to derive (24), except that we use (22) in place of (23),

(28) $\qquad \|\hat{u}(\sigma) - \hat{u}(\tau)\|_{L^2(S^{n-1})} \leq C I_0^{\gamma/2} |\log \tau|^{-\gamma/2}$

for all $0 < \sigma < \tau \leq \tfrac{1}{8}$, and again by Corollary 1 of Section 4.4 and (vii) we conclude that

(29)
$\quad \text{sing } \tilde{u} \cap B_\rho(z) \subset$ the $(\zeta\rho)$-neighbourhood of $z + \{0\} \times \mathbb{R}^{n-3}$ for all $\rho \in (0, \tfrac{1}{2}]$.

In view of (15), (25), (26), and (29) it is now evident that, provided (20) holds, we can take the collection $\{B_{d(z_{j,k})/4}(z_{j,k})\}_{z_{j,k} \in \mathcal{E}_0} \cup \{B_{\sigma_k}(z_k)\}$ to be the collection corresponding to $\mathcal{F}_0$ in the rectifiability lemma of Section 4.2 in case we use ζ in place of ε, and then hypothesis I(b) of that lemma is satisfied in case $x_0 = 0$ and $\rho_0 = 1$.

On the other hand if (20) fails then some ball $B_{1/4}(y)$ with $y \in B_{5/8}(0) \cap S_+$ must have a $\tfrac{\delta}{2}$-gap, and so the first alternative hypothesis in (I) of the rectifiability lemma holds in case $x_0 = 0$ and $\rho_0 = 1$.

Thus, provided ε is sufficiently small (depending on δ, n, N, β), we have shown that $S_+(w_1, \sigma_1)$ (as defined in (vi)) satisfies the hypotheses of the rectifiability lemma of Section 4.2 for $x_0 = 0$, $\rho_0 = 1$. That is, in view of the arbitrariness of w_1, σ_1, we have shown that $S = \overline{B}_{\varepsilon\sigma_0/4}(w_0) \cap S_+$ satisfy the hypotheses I, II of the rectifiability lemma, where $\rho_0 = \frac{\varepsilon\sigma_0}{4}$.

Thus the rectifiability lemma implies that $\overline{B}_{\varepsilon\sigma_0/4}(w_0) \cap S_+$ is $(n-3)$-rectifiable.

Finally, let B be any closed ball contained in Ω. Then by monotonicity (see Section 2.4) there is a fixed $\beta > 0$ such that $\Theta_u(y) \leq \beta$ for each $y \in B$. In particular

$\Theta_\varphi(0) \leq \beta$ for any tangent map of u at any point $y \in B$, and by Lemma 1 of Section 4.4 we know that $\{\Theta_u(y) : y \in \mathrm{sing}_* u \cap B\}$ is a finite set $\alpha_1 < \cdots < \alpha_N$ of positive numbers, where $\mathrm{sing}_* u$ is as in Section 3.5. Let

$$S_j = \{z \in \mathrm{sing}\, u : \Theta_u(z) = \alpha_j\}$$
$$S_j^+ = \{z \in \mathrm{sing}\, u : \Theta_u(z) \geq \alpha_j\}.$$

Notice that S_j^+ is closed in Ω by the upper semi-continuity (ii) of Section 2.5 of Θ_u. Take any $j \in \{1, \ldots, N\}$ and any $y \in S_j$. According to the above discussion, there is $\rho > 0$ such that $B_\rho(y) \cap S_j^+$ is $(n-3)$-rectifiable. Thus, in view of the arbitrariness of y, the set S_j has an open neighbourhood U_j such that

(30) $\qquad\qquad S_j^+ \cap U_j$ is locally $(n-3)$-rectifiable.

Of course the $S_j^+ \cap U_j$ are also locally compact, because S_j^+ is closed and U_j is open. Now let

$$V_j = \{z \in \mathrm{sing}\, u : \Theta_u(z) < \alpha_{j+1}\}, \quad j = 0, \ldots, N-1, \quad V_N = \Omega.$$

Then the V_j are open in Ω by the upper semi-continuity (ii) of Section 2.5 of Θ_u. Now, with $\alpha_0 = 0$, $\alpha_{N+1} = \infty$, $S_0^+ = \mathrm{sing}\, u$, and $U_0 = \emptyset$, we can write

$$B \cap \mathrm{sing}\, u = \bigcup_{j=0}^{N} \{z \in B \cap \mathrm{sing}\, u : \alpha_j \leq \Theta_u(z) < \alpha_{j+1}\}$$
$$= \bigcup_{j=0}^{N} B \cap S_j^+ \cap V_j$$
$$= \left(\bigcup_{j=0}^{N} (B \cap S_j^+ \cap U_j \cap V_j)\right) \cup \left(\bigcup_{j=0}^{N} (B \cap S_j^+ \setminus U_j) \cap V_j\right).$$

This is evidently a decomposition of $B \cap \mathrm{sing}\, u$ into a finite union of pairwise disjoint locally compact sets, each of which is locally $(n-3)$-rectifiable; in fact for each j the set $(B \cap S_j^+ \setminus U_j) \cap V_j \subset \mathrm{sing}\, u \setminus \mathrm{sing}_* u$, and hence has Hausdorff dimension $\leq n-4$ by Corollary 1 of Section 3.4, and the set $B \cap S_j^+ \cap U_j \cap V_j$ is locally $(n-3)$-rectifiable by (30). This completes the proof of Theorem 2. $\qquad\square$

Proof of Remark (2) of Section 4.1:

We have to show that for $\mathcal{H}^{n-3}$-a.e. $z \in \mathrm{sing}\, u$ there is a unique tangent space for $\mathrm{sing}\, u$ at z in the Hausdorff distance sense, and also that u has a unique tangent map at z.

For the former of these we have to show that, for $\mathcal{H}^{n-3}$-a.e. $z \in \mathrm{sing}\, u$, there is an $(n-3)$-dimensional subspace L_z such that for each $\varepsilon > 0$

(1) $\qquad\qquad B_1(0) \cap \eta_{z,\sigma}(\mathrm{sing}\, u) \subset$ the ε-neighbourhood of L_z

4.8. The case when Ω has arbitrary Riemannian metric

and

(2) $\qquad B_1(0) \cap L_z \subset$ the ε-neighbourhood of $\eta_{z,\sigma}(\text{sing } u)$

for all $\sigma \in (0, \sigma_0)$ where $\sigma_0 = \sigma_0(\varepsilon, u, z) \downarrow 0$ as $\varepsilon \downarrow 0$. Using the notation in the last part of the proof above, let $z \in S_j$ be any point where S_j^+ has an approximate tangent space. Thus there is an $(n-3)$-dimensional subspace L_z with

(3) $\qquad \lim_{\sigma \downarrow 0} \int_{\eta_{z,\sigma}(S_j^+)} f \, d\mathcal{H}^{n-3} = \int_{L_z} f \, d\mathcal{H}^{n-3} \quad \forall f \in C_c^0(\mathbb{R}^n).$

We show that (1) and (2) hold with this L_z. In fact the inclusion (2) is evidently already implied by this, so we need only prove (1). Let $\sigma_k \downarrow 0$ be arbitrary, and let φ be any tangent map of u at z with $u(z + \sigma_{k'} x) \to \varphi$ for some subsequence $\sigma_{k'}$. By (3) we evidently have that the ε_k neighbourhood of $B_1(0) \cap \eta_{z,\sigma_{k'}} S_j^+$ contains all of $L_z \cap B_{1/2}(0)$ for some sequence $\varepsilon_k \downarrow 0$. But then by the upper semi-continuity as in Section 2.11 we have

$$\Theta_\varphi(y) \geq \Theta_\varphi(0) = \Theta_u(0) \quad \text{everywhere on } L_z \cap B_{1/2}(0).$$

Thus by (i), (ii) of Section 3.3 we have $S(\varphi) \supset L_z$, and since L_z has maximal dimension $n-3$, this shows that $S(\varphi) = L_z$, so φ is an HCM with $S(\varphi) = L_z$. But then by Corollary 1 of Section 4.4 we have

$$B_1(0) \cap \eta_{z,\sigma_{k'}}(\text{sing } u) \subset \text{the } \varepsilon_k\text{-neighbourhood of } L_z$$

for some sequence $\varepsilon_k \downarrow 0$. In view of the arbitrariness of the original sequence σ_k we thus have (2) as claimed.

Finally we want to show that there is a unique tangent map of u at $\mathcal{H}^{n-3}$-a.e. $z \in \text{sing } u$. Let $S_j = \{z \in \text{sing } u : \Theta_u(z) = \alpha_j\}$ as above. For each $\varepsilon > 0$, we can subdivide S_j into $\cup_{i=1}^\infty S_{j,i}$, where $S_{j,i}$ denotes the set of points $z \in S_j$ such that the conclusions (1) and (2) hold with $\sigma_0 = \frac{1}{i}$. Provided the original w_0, σ_1 in the definition (iv) of $\tilde{u}$ are selected with $w_0 \in S_{j,i}$ and $\sigma_1 = \sigma_1(\varepsilon, u, w_0, i) \leq \frac{1}{i}$, we then have by (1) and (2) that *all* points of $z \in \eta_{w_0, \sigma_1} S_{j,i}$ are contained in the set T_0^+ in the proof of Theorem 2 above. Hence by (28) of the above proof we conclude that there is a unique tangent map of u at each point $z \in S_{j,i} \cap B_{\sigma_1}(w_0)$ with the exception of a set of $\mathcal{H}^{n-3}$-measure $\leq \varepsilon \sigma_1^{n-3}$. In view of the arbitrariness of ε, w_0 here (and keeping in mind that we have already established that $S_{j,i}$ is $(n-3)$-rectifiable) this shows that there is a unique tangent map of u for $\mathcal{H}^{n-3}$-a.e. points $z \in S_{j,i}$. Since $\mathcal{H}^{n-3}(\text{sing } u \setminus (\cup_{i,j} S_{j,i})) = 0$, this completes the proof. $\square$

4.8 The case when Ω has arbitrary Riemannian metric

So far we gave the proof of Theorem 2 for the case when Ω has the standard Euclidean metric. The changes needed in the above arguments to handle the case when Ω is equipped with an arbitrary smooth Riemannian metric

$$\text{(i)} \qquad \sum_{i,j=1}^{n} g_{ij}(x)\, dx^i \otimes dx^j, \quad g_{ij} \in C^\infty(\Omega),\ (g_{ij}) > 0,$$

are of a purely routine technical nature, and we wish to describe them here.

In view of the local nature of the claim of Theorem 2, there is no loss of generality in assuming from the outset that Ω is bounded, $g_{ij} \in C^3(\overline{\Omega})$, and (rescaling the metric by a suitable constant if necessary)

$$\text{(ii)} \qquad |g_{ij}|_{C^3(\Omega)} \le \beta_1, \quad \sum_{i,j=1}^n g_{ij}\xi^i\xi^j \ge |\xi|^2,$$

where β_1 is a fixed constant. Now with the metric (g_{ij}) in place of the Euclidean metric, the energy of $u \in W^{1,2}(\Omega; N)$ is defined by

$$\text{(iii)} \qquad \mathcal{E}^{(g)}(u) = \int_\Omega |Du|_g^2 \sqrt{g}\, dx,$$

where $|Du|_g^2 = \sum_{i,j=1}^n g^{ij} D_i u \cdot D_j u$, with $(g^{ij}) = (g_{ij})^{-1}$ and $g = \det(g_{ij})$. The Euler-Lagrange system (corresponding to (iii) of Section 2.2) for a minimizer or stationary value u is

$$\text{(iv)} \qquad \Delta_g u + \sum_{i,j=1}^n g^{ij} A_u(D_i u, D_j u) = 0,$$

where $\Delta_g u = (\Delta_g u^1, \ldots, \Delta_g u^p)$, with $\Delta_g = g^{-1/2} \sum_{i,j=1}^n D_i(\sqrt{g} g^{ij} D_j)$ the Laplace-Beltrami operator relative to the metric (g_{ij}), and corresponding to the identity (iv) of Section 2.2 we have

$$\text{(v)} \qquad \int_\Omega \left(\sum_{i,j=1}^n (|Du|_g^2 \delta_{ij} - 2 \sum_\ell g^{\ell i} D_\ell u \cdot D_j u) D_i \zeta^j + \sum_{j=1}^n R_j \zeta^j \right) = 0,$$

for any $\zeta = (\zeta^1, \ldots, \zeta^n) \in C_c^\infty(\Omega; \mathbb{R}^n)$ where $|R_j| \le C|Du|^2$ at each point of Ω.

Now let $u \in W^{1,2}_{\text{loc}}(\Omega; N)$ be any minimizer for $\mathcal{E}^{(g)}$, and let $z \in \Omega$. Let $T_z : \mathbb{R}^n \to \mathbb{R}^n$ be an affine transformation with $T_z(z) = z$ and with $\widetilde{T}_z \equiv T_z - z$ satisfying

$$\widetilde{T}_z^t(g_{ij}(z))\widetilde{T}_z = (\delta_{ij}),$$

and with T_z depending continuously on z. Then for each fixed $z \in \Omega$ we evidently have that $u^{(z)} = u \circ T_z$ is $\mathcal{E}^{g^{(z)}}$ minimizing, where

$$(g_{ij}^{(z)}(x,y)) = \widetilde{T}_z^t(g_{ij}(x,y))\widetilde{T}_z$$

hence in particular

$$g_{ij}^{(z)}(z) = \delta_{ij}, \quad z \in \Omega.$$

4.8. The case when Ω has arbitrary Riemannian metric

Then by the appropriate minor modifications of the arguments in Section 2.4 using the identity (v) with $g_{ij}^{(z)}$ in place of g_{ij} instead of the identity (iv) of Section 2.2, we obtain

(vi)
$$\int_{\Omega^{(z)}} (|Du^{(z)}|^2 \delta_{ij} - 2D_i u^{(z)} \cdot D_j u^{(z)}) D_i \zeta^j = \int_{\Omega^{(z)}} (\sum_{j=1}^n S_j \zeta^j + \sum_{i,j} S_j^i D_i \zeta^j)$$

for any $\zeta \in C_c^\infty(\Omega^{(z)})$, where

(vii) $\qquad u^{(z)} = u \circ T_z, \quad \Omega^{(z)} = T_z^{-1}(\Omega), \quad |S_i^j| + \rho |S_j| \leq C\beta_1 |Du^{(z)}|^2.$

It is then easy to see that in place of (ii) of Section 2.4 we obtain

$$(n-2-C\beta_1\rho)\int_{B_\rho(z)} |Du^{(z)}|^2 \leq (1+C\beta_1\rho)\rho \int_{\partial B_\rho(z)} |Du^{(z)}|^2 - 2\rho \int_{\partial B_\rho(z)} |u_{R_z}^{(z)}|^2,$$

and in place of (iii) of Section 2.4.

(viii) $\qquad e^{C\beta_1\rho}\rho^{2-n}\int_{B_\rho(z)} |Du^{(z)}|^2$ is an increasing function of ρ,

so the limit

(ix) $\qquad \Theta_u(z) \equiv \lim_{\rho\downarrow 0} \rho^{2-n} \int_{B_\rho(z)} |Du^{(z)}|^2$

exists at every point $z \in \Omega$. Θ_u is as before called the density function of u. Also, in place of (iii) of Section 2.5 we obtain the inequality

(x) $\qquad 2\int_{B_\rho(z)} \frac{|R_z u_{R_z}^{(z)}|^2}{R_z^n} \leq \rho^{2-n}\int_{B_\rho(z)} |Du^{(z)}|^2 - \Theta_u(z) + E, \quad |E| \leq C\beta_1\rho,$

where $C = C(n, N, \beta)$, β any constant such that $\rho^{2-n}\int_{B_\rho(z)} |Du^{(z)}|^2 \leq \beta$. It follows that we can take tangent maps and pseudo-tangent maps by exactly the same procedure that we used before; thus for example for each $z \in \operatorname{sing} u$ and each $\sigma_j \downarrow 0$, there is a subsequence $\sigma_{j'}$ such that $u_{z,\sigma_{j'}}^{(z)} \to \varphi$, where $u_{z,\sigma}^{(z)}(x,y) = u^{(z)}(z + \sigma(x,y))$ and where $\varphi \in W_{\text{loc}}^{1,2}(\mathbb{R}^n; N)$ is energy minimizing (with respect to the standard metric for $\mathbb{R}^n$) and is homogeneous of degree zero.

Also all of the energy estimates and L^2 estimates of Section 4.4 and 4.5 carry over to the present setting with only very minor changes to take account of the fact that in place of the standard metric (δ_{ij}) we now have (near each point $z \in \operatorname{sing} u$) the metric $g_{ij}^{(z)}$ which satisfies $|g_{ij}^{(z)} - \delta_{ij}| \leq C\beta_1\rho$ on any ball $B_\rho(z)$. Thus for example Theorem 1 of Section 4.1 continues to hold, except that in place of the main inequality we now have

(xi)
$$\rho^{-n}\int_{B_{\rho/2}^+} \left| \int_{S^2} r^2(|Du^{(0)}|^2 - |D\varphi|^2) \, d\omega \right| r^2 \, dr \, dy \leq C\rho^{-n}\int_{B_\rho(0)} r^2((u_r^{(0)})^2 + (u_y^{(0)})^2) +$$
$$+ C\left(\rho^{-n}\int_{B_\rho(0)\setminus\{(x,y):|x|<\rho/2\}} r^2((u_r^{(0)})^2 + (u_y^{(0)})^2)\right)^{1/(2-\alpha)} + E,$$

where $u^{(0)}$ means $u^{(z)}$ with $z = 0$ and where $|E| \leq C\beta_1\rho$. Likewise the main estimate of Theorem 1 of Section 4.5 becomes

(xii)
$$\int_{B_{\rho/4}(z_0)} \frac{|R_z u_{R_z}^{(z)}|^2}{R_z^n} \leq C\rho^{-2} s^{2-n}(z) \int_{B_{3\rho/2}(z_0)} \sum_{j=0}^{n-3} |R_{z_j} u_{R_{z_j}}^{(z)}|^2 + C \left(\frac{\rho^{n-2}}{s^{n-2}(z)}\right)^{1-1/(2-\alpha)} \times$$
$$\left(\int_{B_{3\rho/2}(z_0)\setminus\{(x,y):|x-\xi_{z_0}|<\rho/8\}} \left(\frac{1}{\rho^2 s^{n-2}(z)} \sum_{j=0}^{n-3} |R_{z_j} u_{R_{z_j}}^{(z)}|^2 + \frac{|R_z u_{R_z}^{(z)}|^2}{R_z^n}\right)\right)^{1/(2-\alpha)} + E,$$

for $z \in S_+ \equiv \{w \in \overline{B}_1(0) : \Theta_u(z) \geq \Theta_u(0)\}$, where $u^{(z)}$ is as above, where we are assuming that $\overline{B}_1(0) \subset \Omega$ with metric g still satisfying (ii), and where $|E| \leq C\beta_1\rho^{1+n-2}/s^{n-2}(z)$.

In view of these estimates it is clear that we can take in place of the deviation function ψ of (ii) the function

(xiii)
$$\widetilde{\psi}(x,y) = \int_{S_+} \frac{|R_z u_{R_z}^{(z)}|^2}{R_z^n}\bigg|_{(x,y)} d\mu^+(z).$$

Then by very straightforward modifications of Section 4.6 (using (xi), (xii) in place of the main results in Theorem 1 of Section 4.4 and Theorem 1 of Section 4.5) we can prove the result

(xiv)
$$I_{\theta\rho} \leq C(I_\rho - I_{\theta\rho})^{1/(2-\alpha)}$$

analogous to Theorem 1, where now C also depends on β_1, and where $I_\rho = \int_{T_\rho^+} \widetilde{\psi} + \beta_1\rho$. Using the same kind of iteration as in Section 4.7 (based on (xiv) and on (ix) of Section 4.7), this leads directly to

$$\int_{T_\rho^+} |\log d|^{1+\gamma} \widetilde{\psi} \leq C I_{1/4}^\gamma$$

for $\rho \leq \frac{1}{4}$, where d is defined as in Section 4.7.

The rest of the proof is completed as in Section 4.7, applying all of the above with the rescaled function u_{w_1,σ_1} as in (iv) of Section 4.7 in place of u; this rescaled function lives on the rescaled domain where the appropriately scaled metric satisfies (ii) with the same fixed constant β_1 independent of σ_1, provided we always take $\sigma_1 \leq 1$, which of course we can do—indeed it is necessary in the argument of Section 4.7 only that σ_1 is sufficiently small. We also need to take σ_1 small enough so that the transformation T_z in (vii) satisfies $\|T_{z_1} - T_{z_2}\| \leq \varepsilon$ for $z_1, z_2 \in B_{\sigma_1}(w_0)$. (This of course can be done because T_z is continuous in z.) Then for small enough $\varepsilon = \varepsilon(n, N, \beta)$ we again conclude as we did in Section 4.7 that the hypotheses of the rectifiability lemma of Section 4.2 hold, and hence that $S_+ \cap \overline{B}_\sigma(w_0)$ is $(n-3)$-rectifiable for small enough $\sigma = \sigma(w_0, u) > 0$ and for any $w_0 \in \text{sing}_* u = \{z \in \text{sing } u : \Theta_u(z) = \Theta_\varphi(0) \text{ for some HCM } \varphi\}$. The proof is then completed as before. $\square$

Bibliography

[Ad75] R. A. Adams: Sobolev spaces. Academic Press (1975)

[Al72] W. Allard: On the first variation of a varifold. Ann. of Math. **95** (1972), 417–491

[Ag83] F. Almgren: Q-valued functions minimizing Dirichlet's integral and the regularity of area minimizing rectifiable currents up to codimension two. Bull. Amer. Math. Soc. (N.S.) **8** (1983), 327–328

[AS88] D. R. Adams, L. Simon: Rates of asymptotic convergence near isolated singularities of geometric extrema. Indiana Univ. Math. J. **37** (1988), 225–254

[DeG61] E. De Giorgi: Frontiere orientate di misura minima. Sem. Mat. Scuola Norm. Sup. Pisa (1961), 1–56

[FH69] H. Federer: Geometric Measure Theory. Springer-Verlag, Berlin–Heidelberg–New York (1969)

[FH70] H. Federer: The singular sets of area minimizing rectifiable currents with codimension one and of area minimizing flat chains modulo two with arbitrary codimension. Bull. Amer. Math. Soc. **76** (1970), 767–771

[Gi84] M. Giaquinta, E. Giusti: The singular set of the minima of certain quadratic functionals. Ann. Scuola Norm. Sup. Pisa **11** (1984), 45–55

[GW89] R. Gulliver, B. White: The rate of convergence of a harmonic map at a singular point. Math. Ann. **283** (1989), 539–549

[GT83] D. Gilbarg, N. Trudinger: Elliptic partial differential equations of second order (2nd ed.). Springer-Verlag, Berlin–Heidelberg–New York (1983)

[HL87] R. Hardt, F.-H. Lin: Mappings minimizing the L^p norm of the gradient. Comm. Pure Appl. Math. **40** (1987), 555–588

[Ha85] M. Hata: On the structure of self similar sets. Japan J. Appl. Math. **2** (1985), 381–414

[He91] **F. Hélein:** Régularité des applications faiblement harmoniques entre une surface et une varitée Riemannienne. C.R. Acad. Sci. Paris Sér. I Math. **312** (1991), 591–596

[Hu81] **J. E. Hutchinson:** Fractals and self similarity. Indiana Univ. Math. J. **30** (1981), 713–747

[Jo84] **J. Jost:** Harmonic maps between Riemannian manifolds. Proc. Centre Math. Anal. Austral. Nat. Univ. **3** (1984)

[Lo65] **S. Łojasiewicz:** Ensembles Semi-analytiques, IHES lecture notes (1965)

[Lu88] **S. Luckhaus:** Partial Hölder continuity for minima of certain energies among maps into a Riemannian manifold. Indiana Univ. Math. J. **37** (1988), 349–367

[Lu93] **S. Luckhaus:** Convergence of Minimizers for the p-Dirichlet Integral. Math. Z. **213** (1993), 449–456

[Ma83] **V. G. Maz'ja:** Sobolev spaces. Springer-Verlag, Berlin a.o. (1983)

[Ms53] **J. M. Marstrand:** Some fundamental geometrical properties of plane sets of fractional dimension. Proc. London Math. Soc. **3** (1954), 257–301

[Mt75] **P. Mattila:** Hausdorff dimension, orthogonal projections and intersections with planes. Ann. Acad. Sci. Fenn. Ser. A I. Math. **1** (1975), 227–244

[Mt82] **P. Mattila:** On the structure of self-similar fractals. Ann. Acad. Sci. Fenn. Ser. A I. Math. **7** (1982), 189–195

[Mt84] **P. Mattila:** Hausdorff dimension and capacities of intersections of sets in n-space. Acta Math. **152** (1984), 77–105

[Mo66] **C. B. Morrey, Jr.:** Multiple integrals in the calculus of variations Grundlehren d. math. Wissenschaften in Einzeldarst., vol. 130. Springer-Verlag, Berlin–Heidelberg–New York (1966)

[Re60] **R. E. Reifenberg:** Solution of the Plateau problem for m-dimensional surfaces of varying topological type. Acta Math. **104** (1960), 1–92

[Riv92] **T. Riviere:** Applications harmoniques de B^3 dans S^2 partout discontinues. [Everywhere discontinuous harmonic mappings from B^3 into S^2.] Seminaire sur les Equations aux Derivees Partielles, Ecole Polytech., Palaiseau, Exp. No. XIX (1992)

[SS81] **R. Schoen, L. Simon:** Regularity of stable minimal hypersurfaces. Comm. Pure Appl. Math. **34** (1981), 741–797

[SU82] **R. Schoen, K. Uhlenbeck:** A regularity theory for harmonic maps. J. Differential Geom. **17** (1982), 307–336

Bibliography

[Si83a] L. Simon: Lectures on Geometric Measure Theory. Proc. Centre Math. Anal. Austral. Nat. Univ. **3** (1983)

[Si83b] L. Simon: Asymptotics for a class of non-linear evolution equations, with applications to geometric problems. Ann. of Math. **118** (1983), 525–572

[Si93] L. Simon: Cylindrical tangent cones and the singular set of minimal submanifolds. J. Differential Geom. **38** (1993), 585–652

[Si92] L. Simon: Singularities of geometric variational problems. Summer school lectures delivered at RGI, Park City, Utah (1992); to appear in the AMS Park City Geometry Series.

[Si] L. Simon: On the singularities of harmonic maps. (To appear)

[Wh92] B. White: Nonunique tangent maps at isolated singularities of harmonic maps. Bull. Amer. Math. Soc. (N.S.) **26** (1992), 125–129

Index

Symbols

Here we include a list of the mathematical symbols we used and the corresponding page number of the first occurrence in the text.

$\mathcal{C}_\beta$, 111
B_ρ^+, $B_\rho^+(y_0)$, 110
$L_i^\perp$, 61
$L_{z,\rho}$, 121
$R_y(x)$, 53
$S(\varphi)$, 53
T_ρ, 105
$W^{1,2}$, 37
$W^{1,2}(S^{n-1}; \mathbb{R}^p)$, 25
$W^{1,2}(S^{n-1} \times [a,b]; \mathbb{R}^p)$, 25
$W^{1,2}(\Omega)$, 5
$W^{1,2}_{\text{loc}}(\Omega; N)$, 19
$[\![z_{k,j}]\!]$, 106
$[u]_{\alpha;\Omega}$, 1
Θ_u, 24
$\Theta_u(y)$, 37
$\|\cdot\|_{W^{1,2}(\Omega)}$, 5
$\eta_{z,\sigma}(x)$, 91
$\lambda_{y,\rho}$, 3
$\nabla_{r,y}$, 112
osc_Ω, 2
$\text{sing}_* u$, 58
$\text{reg } u$, 35
$\text{sing } u$, 35
ψ (deviation function), 129
$\llcorner$, 106
ρ_0-uniform approximation property, 63
ρ_z, 104
ε-Regularity Theorem, 22
m-rectifiable, locally, 91
$u(r,y)$, 115

$u_{y,\rho}(x)$, 51
$\mathcal{E}_{B_\rho(y)}(u)$, 19
$\mathcal{H}^{n-2}$, 36
$\mathcal{S}_j$, 54
$\mathbb{R}P^2$, 92

Citations

This citation list is a cross reference indicating the code [number] of articles and books corresponding to the bibliography and the numbers of the pages where the entries are cited.

[AS88], 67, 69
[Ad75], 6
[Ag83], 54
[FH69], 54, 117
[GT83], 6, 7, 9, 11, 13, 37, 49, 85
[GW89], 67
[HL87], 32
[Ha85], 66
[He91], 22
[Hu81], 66
[Jo84], 55
[Lo65], 69
[Lu88], 25, 33
[Lu93], 25
[Ma83], 6
[Ms53], 66
[Mt75], 66
[Mt82], 66
[Mt84], 66
[Mo66], 37
[Riv92], 22
[SU82], 25, 32
[Si93], 92
[Si92], 92

INDEX

[Si], 92
[Si83a], 54, 117
[Si83b], 67, 82
[Wh92], 51, 67

Keywords

absolutely continuous function, 37
approximate tangent space, 91
approximation property, 57
approximation properties, 62
approximation with harmonic functions, 10

blow up Lemma, 10
blowing up techniques, 92
Borel measures, 104

Campanato, Lemma of, 2
compact, locally, 91
Compactness Lemma, 32
compactness Lemma of Rellich, 6
convex subadditive function Lemma, 31
Corollary of Luckhaus, 27
countably j-rectifiable, 61
cover, Whitney, 117
covering lemma, 94
covering lemma, five times, 117
cylindrical map, homogeneous-, 110

δ-approximation property, 57
δ-gap, 93
δ-radius, 104
δ-tilt, 104
Definition of regular set, 35
Definition of singular set, 35
density function, 24
deviation function ψ, 129
differentiability Theorem, 13
Dirac mass, 106

ε-Regularity Theorem, 22
energy, 19
energy estimates, 110
energy minimizing, 20
estimates of the energy, 110

estimates, L^2-, 120
estimates, a-priori, 11
extension, 6

first variation, 20
five times covering lemma, 117

gap measures, 104
gap, δ-gap, 93
general rectifiability lemma, 92
gradient in L^2, 5

Hölder coefficient, 1
Hölder continuity, 1
harmonic approximation Lemma, 10
harmonic approximation Lemma, rescaled, 14
harmonic functions, 8
harmonic map, stationary, 22
harmonic map, weakly, 22
Hausdorff dimension, 54
Hausdorff distance, 92
Hausdorff measure, 36
HCM, 110
homogeneous cylindrical map, 110
homogeneous degree zero extension, 25
homogeneous degree zero minimizers, 52

inequality of Poincaré, 7
integrability condition, 79, 82, 92
interior Schauder estimates, 11

Koch curve, 66

L^2 estimates, 120
L^2 gradient, 5
Lemma for convex subadditive functions, 31
Lemma of Campanato, 2
Lemma of harmonic approximation, 10
Lemma of Luckhaus, 26
Lemma of Morrey, 8
Lemma of Poincaré, 7
Lemma of Rellich, 6
Lemma of Weyl, 10
Lemma, blow up, 10
Lemma, compactness, 32

Lemma, covering, 94
Lemma, covering five times, 117
Lemma, harmonic approximation
  (rescaled), 14
Lemma, rectifiability, 92
Lemma, technical, 13
Lipschitz continuity, 1
Lipschitz domain, 5
local energy, 19
local Hölder continuity, 1
locally m-rectifiable, 91
locally compact, 91
Luckhaus compactness Lemma, 32
Luckhaus Corollary, 27
Luckhaus, lemma of, 26

main theorems, 91
mean-value property, 9
minimizing energy, 20
mollifier, 4
monotonicity formula, 23
Morrey's Lemma, 8
multi-index, 4

nearest point projection, 20

oscillation, 2

Poincaré's inequality, 7
projection, nearest point, 20
purely 1-unrectifiable, 66

radius, δ-radius, 104
rectifiability lemma, 92
Rectifiability of the singular set, 91
rectifiable, 91
rectifiable, j-rectifiable, 61
rectifiable, m-rectifiable, 91
rectifiable, locally m-, 91
regular set, 35
regularity estimates, 11
Regularity Theorem, 22
Rellich compactness Lemma, 6
rescaled harmonic approximation
  Lemma, 14

Schoen-Uhlenbeck Regularity Theorem,
  22

self-similar set, 66
singular set, 35, 54
slicing by radial distance function, 26
smoothing, 4
Sobolev space, 5, 19
stationary harmonic map, 22

tangent map, 51
tangent space, approximate, 91
technical Lemma, 13
Theorem, ε-Regularity, 22
tilt, δ-tilt, 104
top dimensional part, 58

uniform approximation property, 63
unique tangent maps, 61
unrectifiable, 66
upper semi-continuity, 25
upper semicontinuity, 37

variational equations, 20

weak j-dimensional δ-approximation
  property, 63
weak derivative, 5
weakly harmonic functions, 10
weakly harmonic map, 22
Weyl's Lemma, 10
Whitney cover, 117